IEE TELECOMMUNICATIONS SERIES 32

Series Editors: Professor J. E. Flood
Professor C. J. Hughes
Professor J. D. Parsons

Analogue

OPTICAL

FIBRE

communications

Other volumes in this series:

Analogue
OPTICAL
FIBRE
communications

LIVERPOOL
UNIVERSITY
LIBRARY

FIAT LVX

Editors
B. Wilson, Z. Ghassemlooy
and I. Darwazeh

The Institution of Electrical Engineers

Published by: The Institution of Electrical Engineers, London,
United Kingdom

© 1995: The Institution of Electrical Engineers

The Institution of Electrical Engineers,
Michael Faraday House,
Six Hills Way, Stevenage,
Herts. SG1 2AY, United Kingdom

British Library Cataloguing in Publication Data

A CIP catalogue record for this book
is available from the British Library

ISBN 0 85296 832 9

Printed in England by Short Run Press Ltd., Exeter

Contents

Contents

10   Optical receiver design for optical fibre SCM systems
P. M. R. S. Moreira, I. Z. Darwazeh and J. J. O'Reilly   257

11   Optical fibre amplifiers and WDM
M. J. O'Mahony   286

Preface

It is inevitable that potential readers will initially express doubts concerning the thinking behind a book on analogue fibre communications in this enlightened digital age of multi-gigabit transmission rates on information superhighways. However, as with many areas of scientific enquiry, initial perceptions change the closer one looks into the problem, or indeed, considers just what constitutes the precise nature of the problem. This book is not an attempt to champion analogue techniques in opposition to digital ones; but rather it seeks to bring together recent thinking on combining analogue and binary methods to explore how such a synthesis may be used to produce improved techniques for future fibre communication systems. Much of the impetus to produce this book arose from a special edition of IEE Proceedings-J on Analogue Optical Fibre Communications published in December 1993, compiled by two of the present editors (Brett Wilson and Zabih Ghassemlooy). A substantial proportion of the material contained in this book is, however, completely new and grew out of subsequent discussions with various colleagues, including Izzat Darwazeh, who joins us as a third editor. We are fortunate that many leading authorities from both academic and industrial backgrounds agreed to contribute their latest material to this volume, with, in some cases, experimental results arriving only shortly before the publishing deadline!

The two main technological issues presently facing fibre system designers are those concerned with bandwidth utilisation and mixed-mode traffic. How are we to utilise the potentially enormous multi-terrahertz bandwidth available from fibre with only the restricted bandwidth of around 30 GHz presently available from intensity modulated/direct detection electronic systems and components? While it is often claimed that fibre bandwidth is "free and unlimited", electrical bandwidth in terminal equipment most certainly is not.

Subcarrier multiplexing (SCM) techniques are maturing as a promising method of delivering broadband mixed-mode services, but are not without their attendant problems. Fortunately, recent developments in doped-fibre amplifiers and improved wavelength-division-multiplexing components and techniques have provided a significant boost in achieving the highly-desirable target of optically transparent networks without the data-rate constrictions of digital regenerators.

Combining fibre, radio and free-space techniques also promises to extend full network access to cover highly mobile users in multi-site office and manufacturing environments as part of what may be described as a global optical ether.

Much of the (non-computer generated) data to be conveyed over both fibre point-to-point links and broadcast distribution networks is analogue in its original source form and is still required in an analogue format after passage through a fibre transmission system. Fibre-to-the-curb or even fibre-to-the-home systems for the near future must therefore be able to handle broadband services such as multi-channel television, video-on-demand and high-definition television that are all presently analogue in nature. However, such systems must have the intrinsic potential for graceful evolution to handle mixed-mode traffic as more of the services transform into digital formats as local processing power increases. Modern fibre instrumentation systems are now being required to transmit analogue data with a bandwidth in the GHz region, a task for which pulse time modulation techniques are proving useful because of their much lower bandwidth overhead penalty compared with analogue to digital conversion and PCM .

We have grouped contributions into three different broad categories: general principles; systems and applications; components and technologies. Early chapters consider many of the basic issues of transmitting analogue-sourced or mixed-mode data over fibre. Subcarrier multiplexing techniques are outlined in chapter 1 with coherent transmission methods receiving attention in chapter 2. Work on modulation formats for fibre video transmission is detailed in chapter 3, while chapter 4 looks at the application of pulse time modulation techniques to fibre systems. Chapter 5 explores the theoretical potential of soliton pulse modulation techniques for analogue transmission. Chapter 6 presents analytical and practical aspects of digital pulse position modulation applied to fibre systems.

Later chapters focus on specific systems and address technological issues related to the implementation of analogue optical systems. In chapter 7 the performance of SCM systems is considered with special emphasis on the effects of non-ideal laser diode behaviour. Chapters 8 and 9 detail the use of fibre systems for the generation and delivery of radio wave and millimetre signals in the tens of GHz regime for mobile systems and cellular network applications. In chapter 10 the emphasis is on circuit noise modelling and minimisation for a tuned SCM optical receiver fabricated as a GaAs MMIC using a commercial foundry process. Finally, chapter 11 summarises the latest developments in optical amplifiers and wavelength division multiplexing components and technologies.

Brett Wilson
Zabih Ghassemlooy
Izzat Darwazeh

Manchester, June 1995

Acknowledgements

This book was published from camera-ready copy produced on a 600 dpi laser jet printer by the editors using Word for Windows to edit material supplied by authors on computer disk. The editors are grateful to Giles Nutkins and Jason Drew for their invaluable assistance in converting a number of the authors' electronic diagrams into a format compatible with our word processing software. We would also like to express our thanks to Fiona MacDonald from the IEE publishing section for patiently answering our numerous questions concerning layout and style and with keeping a weather eye on progress and schedules. The relatively short production timescale has inevitably made demands on the authors regarding submission deadlines. We are indebted to them, since modern scientific literature could not thrive in its present form without a body of highly knowledgeable scientific authors willing to expend considerable time and effort contributing individual chapters to multi-author research volumes for a modest financial return.

The author of chapter 2 (A.J. Seeds) would like to thank the members of his research group for their many contributions to the work described, in particular Dr. Bo Cai who carried out pioneering work on both reverse-biased quantum well lasers and optical FM transmission systems. Aspects of the work were supported by the UK Engineering and Physical Sciences Research Council, the US Air Force Office of Scientific Research and the industrial partners of the UK DTI/SERC LINK WORFNET project.

The author of chapter 8 (D. Wake) wishes to acknowledge the valuable contributions made to the work reported by a number of his colleagues at BT Labs, especially Nigel Walker, Ian Smith and David Moodie. In addition, he would like to acknowledge the contribution made by Claudio Lima of the University of Kent.

Finally, we thank our families and close colleagues for their tolerance and support while the book was under preparation.

The editors

Contributors

Dr J M Arnold, Dept of Electronics & Electrical Engineering, University of Glasgow, UK.

Dr M H Capstick, School of Electronic Engineering and Computer Systems, University of Wales- Bangor, UK.

Prof R A Cryan, Department of Electrical, Electronic Engineering and Physics, University of Northumbria at Newcastle, Newcastle upon Tyne, UK.

Dr I Z Darwazeh, Department of Electrical Engineering and Electronics, UMIST, Manchester, UK.

Prof P A Davies, Optical Communications Group, Electronic Engineering Laboratories, University of Canterbury, Kent, UK.

Dr J M H Elmirghani, Department of Electrical, Electronic Engineering and Physics, University of Northumbria at Newcastle, Newcastle upon Tyne, UK.

Dr Z Ghassemlooy, Communications Group, School of Engineering IT, Sheffield Hallam University, Sheffield, UK.

Dr N J Gomes, Optical Communications Group, Electronic Engineering Laboratories, University of Canterbury, Kent, UK.

Dr D T Lambert, Optical Fibres, Deeside, UK.

Dr P M Lane, Department of Electronic & Electrical Engineering, University College London, UK.

Dr P M R S Moreira, ECP Division, CERN, Geneva, Switzerland.

Prof M O'Mahony, Department of Electronic Systems Engineering, University of Essex, Colchester, UK.

Prof. J J O'Reilly, Department of Electronic & Electrical Engineering, University College London, UK.

Dr H M Salgado, School of Electronic Engineering and Computer Systems, University of Wales-Bangor, UK.

Prof. A J Seeds, Department of Electronic & Electrical Engineering, University College London, UK.

Prof. J M Senior, Department of Electrical & Electronic Engineering, Manchester Metropolitan University, Manchester, UK.

Dr R T Unwin, Division of Electronic and Communication Engineering, School of Engineering, University of Huddersfield, UK.

Dr D Wake, British Telecom Laboratories, Martlesham Heath, Ipswich, UK.

Dr B Wilson, Fibre Communications Research Group, Department of Electrical Engineering & Electronics, UMIST, Manchester, UK.

Chapter 1

Subcarrier multiplexing
in optical communication network

P.A.Davies and N.J.Gomes

1.1 Introduction

A major aim of engineers concerned with modern optical communications systems is to find a way around the "electronic bottleneck". This graphic phrase describes the problem of trying to match the bandwidth of optical communication systems to the rather more restricted bandwidth of electronic systems. The low attenuation windows at 1.3μm and 1.55μm have bandwidths of around 0.1μm. If these optical bandwidths are converted to frequency ranges this is equivalent to a total bandwidth of 30,000GHz. In contrast, in electronic systems even with carrier frequencies of many GHz, bandwidths are limited to a few GHz at most. How can the optical bandwidth be exploited when the electronics driving the optical devices is so limited?

Many approaches to solving this problem have been suggested, including optical time division multiplexing (OTDM) [1] and wavelength division multiplexing (WDM) [2]. In this chapter we will consider a simple and cost-effective alternative approach known as subcarrier multiplexing (SCM) [3,4].

The principle of subcarrier multiplexing is reasonably straightforward and it is an easy technique to employ in practice. An RF or microwave signal (the subcarrier) is used to modulate an optical carrier. This results in an optical spectrum consisting of the original optical carrier plus two sidetones located at $f_o \pm f_s$ where f_o is the frequency of the optical carrier and f_s is the frequency of the RF subcarrier. If the subcarrier is then modulated with data (or video or telephone traffic) then sidebands centred on $f_o + f_s$ and $f_o - f_s$ are produced. Multiple channels can be multiplexed onto the same optical carrier by using multiple subcarriers. At the receiver the channels are demultiplexed by using direct detection and then applying heterodyning and filtering to the resultant RF signal. The process is illustrated in Figure 1.1.

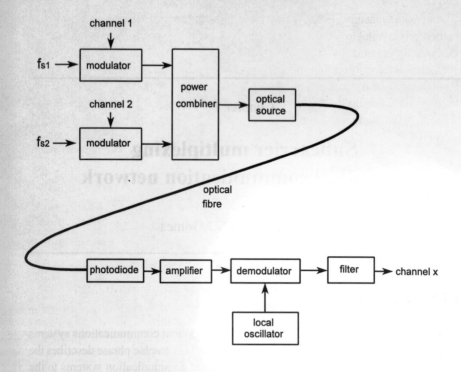

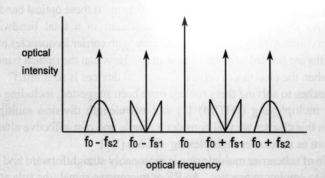

Figure 1.1 *Subcarrier multiplexing and the resulting optical spectrum*

Analogue modulation of optical carriers can be achieved by using external modulators or by direct modulation of semiconductor laser diodes. Present limits to modulation rates are around 20GHz. To use more of the optical bandwidth it is necessary to combine SCM with other more broadband methods of dividing the optical spectrum such as WDM.

A major advantage of SCM is that many of the components required for the system are readily available. Modulators, mixers and amplifiers employed in cable, or community antenna TV (CATV) and satellite TV systems can be used allowing a realistic low-cost solution to the problem of exploiting optical bandwidth. A great strength of SCM is that it can be used for a very wide range of applications. Potential application areas include: microwave and millimetre-wave signal distribution for antenna [6] or fibre radio [7] applications; replacement of coaxial cable systems for cable TV [8]; telephony; data transmission and local area networks (LANs) [9]. It also has the advantage that it is very easy to mix traffic on the same fibre by allocating different subcarriers to different types of traffic. For example analogue video, digital telephony and data transmission can all be carried on the same fibre. This ability to mix analogue and digital transmission is very important for applications such as (optical) cable TV networks since cable TV companies are now permitted to offer additional services such as telephony.

1.2 Noise and distortion in analogue optical communications

Most optical communication systems to date have relied on digital transmission where the effects of noise and distortion are well known. Noise causes bit errors at the decision gate of the receiver and distortion causes changes to the pulse shapes resulting in intersymbol interference, which also produces bit errors. The major parameter, in addition to bandwidth, which characterises a digital optical link is therefore bit error rate (BER). In long links, regeneration of the pulses is often carried out with a slight increase in bit error rate each time the pulses are regenerated.

In analogue optical links the effects of noise and distortion are somewhat different. The signals can be amplified but are not regenerated and therefore any noise or distortion added to the signal as it passes through the system appears at the output of the receiver. Characterisation of the noise and distortion contributions is therefore of great importance and in this section we will consider their effects in analogue optical communication systems.

1.2.1 Noise

In an analogue system the addition of noise degrades the signal-to-noise ratio (SNR). This ratio is of course dependent on bandwidth and may vary with operating frequency but is nevertheless used as a general measure of system performance. In modulated systems the parameter used is carrier-to-noise ratio (CNR). The most demanding requirements are in video applications, particularly when amplitude modulation of subcarriers is used, where CNR >55dB may be required.

There are many sources of noise in an optical communication system. The optical source will usually introduce some unwanted intensity variations, caused by effects such as mode instability or spontaneous emission noise. These intensity noises are usually collected together in the term relative intensity noise (RIN) [10] defined as the ratio of the mean square of the intensity fluctuations to the square of the mean of the intensity. It varies with frequency and for semiconductor laser diodes tends to peak near the relaxation oscillation frequency.

The optical channel is usually fibre and is not considered to add noise to the signal but rather to degrade the signal through attenuation, dispersion and the introduction of distortion through nonlinearities under large signal conditions. Of course, many optical links now include optical amplifiers which will increase the signal level but will also degrade the SNR by at least 3dB.

At the receiver the usual noise sources are added. These include shot (quantum) noise, photodiode dark current noise and thermal noises from the load resistor and from the voltage amplifier. In digital systems the dominant system noise sources tend to be the receiver thermal noise sources and so great efforts are made to reduce these noise contributions. However, in analogue systems the power levels incident on the receiver photodiode are often quite high with a result that the receiver can be operating in the shot-noise-limited domain. The emphasis therefore tends to be on minimising the transmitter noise contribution, i.e. the RIN.

1.2.2 Distortion

In analogue systems distortion is caused by nonlinearities. These can be in the laser intensity-current characteristic or as a result of the raised cosine transfer function of the external modulator. If an optical amplifier is used in the channel then the nonuniform gain profile [11] can also cause distortion of broadband signals. At the receiver nonlinearities in the receiver amplifier can also make a contribution. The effects of these individual contributions are considered later in the chapter; for the present we confine our discussion to the parameters used to quantify distortion in analogue optical communication systems.

Consider the nonlinear transfer function represented by the equation below:

$$V_0 = a V_1 + b V_1^2 + c V_1^3 + \ldots \ldots \tag{1.1}$$

where V_1 is the input signal and V_0 is the output signal. If the input is a signal at frequency f_1 then the higher order terms will generate harmonics at $2f_1$, $3f_1$ etc. The harmonics may fall in-band and cause distortion. If the signal contains more than one frequency, for example two frequencies at f_1 and f_2, then mixed frequency components will occur at f_1+f_2, f_1-f_2, $2f_1 + f_2$, $2f_1 - f_2$ and so on. In a frequency division multiplexed system such as an SCM system some of these components will fall in-band and cause unwanted interference. The term $2f_1$-f_2 commonly occurs and gives rise to the third order intermodulation product (3rd order IMP).

System distortion performance and limits are usually set by collecting all of the appropriate distortion components together into two agreed terms. The composite second order (CSO) and composite triple beat (CTB) levels are the ratios of the sums of the second order components to carrier or third order components to carrier, respectively, that fall in-band. A typical specification for an AM-VSB video distribution system is CSO< -55dBc and CTB< -65dBc.

1.3 Transmitters

In an SCM transmitter the individual subcarriers, modulated in any desired format, are linearly combined, for example in a microwave power combiner, prior to transmission. In an optical SCM system this composite signal is used either to modulate the injection current of a semiconductor laser diode (direct modulation) or to drive an electro-optic modulator (external modulation). With direct modulation both intensity modulation (IM) and frequency modulation (FM) of the optical output generally occur. Photodiode receivers will normally only respond to the intensity modulation but optical frequency modulation can be used in conjunction with optical frequency discriminators [12] or coherent (optical heterodyne) detection techniques [13]. The most common implementations of externally modulated systems employ intensity modulation [14], but phase/frequency modulation can be used [13].

In the following subsections the criteria that need to be considered for optical SCM transmitter design will be described. These include the modulation bandwidths of the transmitter components, the linearity of these components, important in SCM systems to minimise intermodulation distortion, the noise of the sources which is important in obtaining high carrier- and signal-to-noise ratios, and the electro-optic conversion efficiency from the RF/microwave signal to the optical signal which affects overall link loss. Emphasis is placed on IM transmitters as these are currently the most common SCM system implementations.

1.3.1 IM transmitters: direct laser modulation

Figure 1.2 shows the light output power versus injection current (P-I) characteristic of a typical semiconductor laser. The application of a modulation current is seen to cause an output power variation with modulation index dependent on the peak of the input signal and the laser operating point. For high electro-optic conversion efficiency, laser diodes with low threshold currents and steep P-I curves are desirable and both parameters are related to the optical gain in the laser cavity [10]. The use of buried-heterostructure lasers which provide both optical and carrier confinement can provide this gain; such index-guiding structures also aid the realisation of very linear above-threshold characteristics [10]. Quantum-well devices can provide even higher gain and low threshold currents. However, the output power handling capability of the laser is also important in many applications in order to achieve shot-noise-limited detection and it may be better in these cases to tolerate extra overall link insertion loss to ensure this. The availability of optical fibre amplifiers at 1.55μm wavelength may also be a consideration in the choice of source.

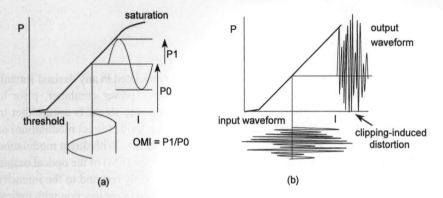

Figure 1.2   *Laser characteristics illustrating (a) optical modulation index (OMI) and (b) clipping-induced distortion*

From the characteristic of Figure 1.2(a) it can be seen that a significant linear region is available for analogue modulation schemes such as SCM. Large optical modulation depths will provide high (sub)carrier- and signal-to-noise ratios. However, if the modulation depth is increased the laser will be driven into its nonlinear regions and distortion will occur. At high power levels the optical output power begins to saturate due to effects such as current spreading and junction heating [10]. Generally, close temperature control will become more problematical and device reliability will be diminished. As discussed later in 1.6.1, shot noise limitations also become more significant for directly modulated links if the average optical power level increases.   A more serious limitation to the signal excursion is the highly nonlinear region around the "knee" of the P-I characteristic near threshold. This clipping-induced distortion (Figure 1.2(b)) imposes a fundamental limitation to directly modulated SCM system performance.

The subcarriers in an SCM system can add to cause large peak deviations. However, since the subcarriers are of different frequencies their peak deviations will only very occasionally add to give a maximum deviation of the composite signal. Rather than use a peak modulation index for the composite signal, an rms modulation index σ is commonly used [15]:

$$\sigma = m \sqrt{\frac{N}{2}} \tag{1.2}$$

where N is the number of channels or subcarriers and m is the modulation index of an individual channel. The fact that the peak excursions of the subcarriers will only occasionally be coincident means that optimum performance is often achieved with some degree of clipping, what is sometimes referred to as "overmodulation"; however, even in such systems σ is less than 1. From Equation 1.2 it is obvious that the modulation index for individual channels must be decreased as the number of subcarrier channels is increased, resulting in a trade-off between CNR and intermodulation distortion. If particular CNR and CSO/CTB requirements are to be met, then given the noise parameters of the system, such as laser RIN and receiver thermal and shot noise, the clipping-induced distortion effectively sets a limit on the number of channels for a given received power level [15,16,17].

When operated well within the linear region of their P-I characteristics, semiconductor lasers can still show a *dynamic* nonlinearity. Although nonlinear gain effects such as spatial and spectral hole-burning, two-photon absorption and lateral carrier diffusion [18] do influence this dynamic nonlinearity, the fundamental cause is the resonant photon-electron interaction within the laser cavity. This resonance distortion is apparent in the IM response of a laser - the peak at the relaxation oscillation (R-O) frequency [10]. The problem can be analysed using direct integration of the laser rate-equations but for subcarrier systems quasi-linear approximations such as Volterra series analyses are adequate. Basically, simulations and experiments show that the modulation frequency, or highest subcarrier frequency in an SCM system, must be significantly lower than the laser R-O frequency if distortion is to be kept low. The precise limitations will be different for different systems applications but if the subcarrier multiplex covers more than an octave of bandwidth the limitations could be especially significant as second-order distortion is high. Darcie and Bodeep [19], for example, examined this problem for NTSC CATV transmissions and indicated that for the CATV CSO (composite second-order products) requirements to be met a 7 GHz resonance frequency laser diode would be required, even though the CATV bandwidth did not extend beyond 550 MHz. This shows that the modulation bandwidth requirements of lasers in directly modulated systems are largely dependent on intermodulation performance and are much more demanding than may initially seem the case. At present, single-longitudinal-mode semiconductor lasers such as DFB (distributed feedback) or DBR (distributed Bragg-reflector) lasers, which are preferred to avoid fibre dispersion problems (see next section), are limited to modulation bandwidths (R-O frequencies) of up to about 20 GHz [20,21].

The associated FM (chirp) that occurs with direct laser modulation can result in dispersion-induced distortion after propagation through an optical fibre (see next section). It is therefore also worth noting that the FM response of a semiconductor laser is enhanced due to thermal effects at low frequencies (below 10MHz) and also peaks at the laser R-O frequency [10].

Finally, the static linewidth of the semiconductor laser needs to be considered. Although narrow linewidth is generally desirable to minimise dispersion, very narrow linewidths can lead to increased problems from reflections from fibre splices, connectors etc. Coherent feedback, which becomes more likely with narrow linewidth (long coherence length) sources, causes modulation of the optical output and distortion [22]. Thus, although DFB and DBR linewidths of a few MHz are readily obtainable, devices with 50 MHz linewidths have been proposed for multichannel applications [21].

1.3.2 IM transmitters: external modulation

Externally modulated optical systems offer a number of advantages deriving from the use of CW (continuous-wave) laser sources. Semiconductor lasers exhibit less RIN when not modulated [23] and, of course, problems such as chirp and distortion in the laser are eliminated. Probably the major attraction of such systems is the ability to use higher power solid-state sources. Nd^{3+}:YAG (neodymium - yttrium aluminium garnet) lasers have been the most popular solid-state sources: these devices operating at 1.32µm wavelength can typically deliver 50 - 200 mW of optical power into a fibre (at least an order of magnitude higher than a semiconductor laser) and have lower RIN, approximately -165 dBc/Hz compared to about -155 dBc/Hz for a DFB laser.

Nd:YAG lasers need to be optically pumped, usually by high power (large optical cavity) semiconductor laser diodes operating at about 800nm wavelength. The diode pump sources will typically have electrical power requirements of a few watts and emit between 500 mW and 1 W optical power for coupling into the Nd:YAG laser. The Nd:YAG output should be optically isolated to minimise the effects of back reflections from the modulator and fibre. These lasers typically have relaxation oscillation frequencies of around 200 kHz. Although this is well below the subcarrier frequencies used in most SCM applications this noise is upconverted around each subcarrier frequency. Nd:YAG lasers for analogue communications systems suppress this noise using either feedback control to the pump source or feedforward control to the modulator [14].

For externally modulated analogue systems the main disadvantage is the nonlinearity of the light output power versus voltage transfer characteristics of the interferometric intensity modulators used. Two related structures are used: the Mach-Zehnder (MZ) modulator and the Y-fed directional coupler modulator.

In the MZ device the input light is split into two arms and then recombined in a single output waveguide to give constructive or destructive interference according to the applied voltage to the device (Figure 1. 3(a)). The transfer characteristic is a raised-cosine function:

$$P_o = P_i t_{ff}[1 + \cos(\frac{\pi V}{V_\pi} + \phi_b)]$$

(1.3)

where t_{ff} is the optical transmission factor of the device, and relates to its excess loss, V is the applied voltage, V_π is the voltage required to vary the light output from a minimum to a maximum (often called the "switching voltage") and ϕ_b is a bias phase (as the lengths in a real device are never exactly matched at zero applied bias).

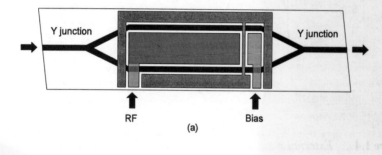

(a)

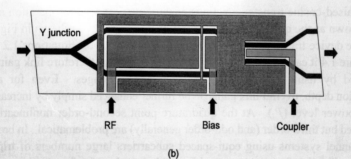

(b)

Figure 1.3 *External intensity modulator structures: (a) Mach Zehnder modulator; (b) directional coupler modulator. Courtesy of Integrated Optical Components*

In the directional coupler type of modulator (Figure 1.3(b)) the combiner section is replaced by a 1×2 directional coupler. Two outputs are available, which are in anti-phase. The transfer characteristics are again raised-cosine functions for each output:

$$P_{o\pm} = P_i t_{ff}[1 \pm \cos(\frac{\pi V}{V_\pi} + \phi_b)] \qquad (1.4)$$

where the symbols are as defined for Equation 1.3 and the plus/minus subscripts denote the two different available outputs. The directional coupler modulator has an inherently lower loss than the MZ type at the expense of a somewhat more complex structure.

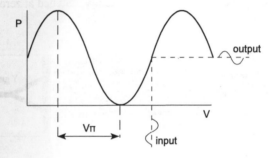

Figure 1.4   *External modulator raised cosine transfer characteristic*

The raised-cosine transfer functions of these devices lead to operation at a bias point known as the quadrature point (the half-power point) as shown in Figure 1.4. From the device transfer functions or P-V characteristics of Equations 1.2 and 1.3 and Figure 1.4 it can be seen that the slope efficiency and therefore link gain can be improved by using devices with low switching voltages. Even for a given modulation depth, overall link gain can be further enhanced simply by increasing the source power level (P_i). At the quadrature point second-order nonlinearities are minimised but third-order (and odd-order generally) are problematical. In broadband multichannel systems using equi-spaced subcarriers large numbers of triple beat products will fall at frequencies coincident with the subcarrier frequencies, especially near the centre of the band [14]. Broadband modulators will typically use travelling-wave electrode designs which attempt to velocity-match the optical and RF waves. However, this requires short electrodes for high-frequency devices; the reduction in interaction length results in larger switching voltages (for the same material / same electro-optic coefficient).

1.3.3 IM transmitters: compensation techniques

Significant effort has been made in developing techniques for the compensation of optical transmitter nonlinearity in SCM systems and analogue optical systems generally. This is especially true for externally modulated systems where the modulator transfer function is a major system limitation but, as will be discussed later, is also the case for directly modulated systems.

External modulator compensation techniques can be divided into two categories: optical techniques and electronic techniques. Probably the most straightforward optical compensation technique is that of polarisation mixing [24]. In LiNbO$_3$ (lithium niobate) MZ modulators the two orthogonal polarisation states in the crystal have different electro-optic coefficients (by a factor of approximately 3). This means that there are two superposed raised-cosine transfer functions of differing periodicity. Normally the polarisation is controlled so that the TE mode (with smaller V_π) is used; however, by allowing a certain amount of the TM mode to propagate, sections of the overall P-V characteristic can be partially linearised, at the expense of a decrease in slope efficiency in the linearised regions. The main disadvantage of this technique is the close control and adjustment of polarisation required which can prove difficult in real systems.

More advanced optical compensation techniques involve modifications of external modulator architectures. For example, series [25] and parallel [26] combinations of Mach-Zehnders have been used. However, such schemes generally suffer from the need to control closely the parameters of two (or more) modulators and to drive the modulators correctly in order to take into account phase/propagation delays (especially for series combinations) which can limit the bandwidth of the compensation. Using directional coupler modulators some compensation has been achieved through small modifications to the "passive" output coupler section; these devices then have the advantage that the RF signal is applied to the first coupler section only and just fixed (optimised) biases are applied to the phase bias and output coupler sections to give compensation.

A wide variety of electronic compensation techniques have been proposed and demonstrated [28,29]. However, two techniques have been particularly extensively investigated and it is these two techniques that have been adopted in commercially available CATV transmitters.

The first of these techniques is predistortion, in which the subcarrier signals to be transmitted are predistorted by feeding them through a nonlinear network with an opposing nonlinearity to that of the transmitter itself. Diode predistortion networks have been most commonly used for externally modulated analogue optical systems as shown in Figure 1.5. Although high levels of compensation can be achieved in this manner, it is very dependent on the signal levels in the system and the bias on the diode predistortion network [30]. Commercial units such as that produced by

Harmonic Lightwaves thus require close feedback control of the predistortion network bias [14]

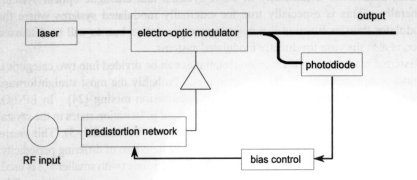

Figure 1. 5   *The use of a diode predistortion network for compensation of intermodulation distortion*

In the feedforward [31] (or quasi-feedforward) technique a fraction of the light output from the modulator is detected, compared with the RF input to the modulator, to generate an error signal and this error signal is fed forward to a second modulator, to a second laser-modulator pair or to a directly modulated laser which compensates in the optical domain. The general scheme is shown in Figure 1.6. If a second laser is used it is important that its wavelength is sufficiently different to that of the original laser in order that coherent interference problems are avoided. A feedforward technique is used in Magnavox CATV transmitters. The main drawback of the technique is the bandwidth limitations that could arise in the electronic circuitry [14]. Also, if a second laser is used, which cannot transmit at too close a wavelength to the original laser, the compensation technique could be limited by fibre dispersion.

With linearised intensity modulators, externally modulated SCM transmitters employing high power solid-state sources lasing in the fibre minimum dispersion window have become very attractive for CATV and other applications. However, the case for directly modulated transmitters has been much strengthened by the now wide availability of fibre amplifiers at the minimum loss window of 1.55µm and, to a lesser extent, by the availability of dispersion-shifted fibre.

For directly modulated systems, simple piecewise linear compander functions have been used to compensate for static nonlinearity [32]. The dynamic nonlinearity, which should not be significant for VHF and UHF applications if high-frequency

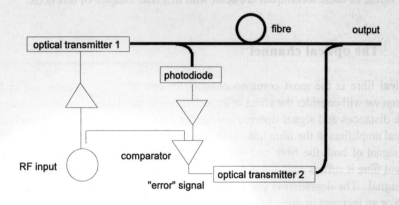

Figure 1. 6   *Feedforward compensation. Note: the optical transmitter can be (a) two directly modulated laser diodes, (b) an externally modulated source for 1, with a directly modulated compensation transmitter for 2, or (c) two external modulators in which case they would be fed from the same optical source*

lasers are used, can also be reduced by using feedforward techniques.   A furtheradvantage that feedforward linearisation can bring to directly modulated systems is a simultaneous cancellation of intensity noise [33], which is a major consideration for these systems. It should also be remembered that optical amplifiers add further noise and distortion and these effects are dealt with in the following section.

1.3.4 High-frequency subcarrier generation

Semiconductor lasers and external modulators are presently limited to modulation rates of up to approximately 20GHz and 40GHz respectively, even using experimental devices [21,34]. It has previously been stated that in multichannel systems, laser modulation rates may be further restricted when resonance distortion is taken into account. For the generation of higher frequency microwave or millimetre-wave subcarriers alternative techniques need to be considered. Device nonlinearities can be used to generate a frequency comb, by mode-locking or gain-switching [35], using self-pulsating lasers [36], or in over-driving external modulators [37]. Spectral techniques include optical heterodyning [38], injection-locking using FM sidebands [39], and FM-IM conversion through dispersive fibre [7]. Some of these techniques are dealt with in a later chapter of this book.

1.4   The optical channel

Optical fibre is the most common channel for optical SCM systems and in this section we will consider the effect of imperfections in the channel on the SCM signal. Link distances and signal distribution can be increased significantly by including optical amplifiers in the fibre link. It is important therefore to consider the effect on the signal of both the fibre and the fibre amplifiers. When light passes down an optical fibre it suffers attenuation and dispersion both of which cause degradation of the signal. The degradation can take the form of a reduction in the signal-to-noise ratio or an increase in distortion.

1.4.1 Attenuation

Attenuation in fibre is relatively low at around 0.5dB/km or less for the 1.3µm and 1.55µm windows; however, it is important that reasonably high signal levels appear at the optical receiver so that the effects of receiver noise may be minimised. In some systems optical amplifiers, usually optical fibre amplifiers, are added to the system to allow greater link lengths or distribution to a larger numbers of users. While optical amplifiers can easily provide gains of 20dB or more they do add noise to the signal, characterised by the noise figure of the amplifier.

The most common optical amplifier at present is the erbium-doped fibre amplifier (EDFA) which operates at 1.55µm. Amplifiers for 1.3µm are not yet generally available, so we will confine ourselves to the EDFA for the present.

There are several contributions to the noise process in an EDFA. In its simplest form this amplifier consists of an optical pump source, operating at one of the usual pump wavelengths, for example 980nm, and a signal source at 1.55µm, coupled into a length of erbium-doped fibre. In the erbium-doped fibre electrons are pumped to a higher energy level and fall to a lower energy level either through spontaneous emission or by stimulated emission at 1.55µm. In the absence of a signal the spontaneously emitted photons act as a signal source and are amplified through stimulated emission over a wide optical bandwidth causing a noise floor known as amplified spontaneous emission (ASE) noise. A typical EDFA output is shown in Figure 1.7.

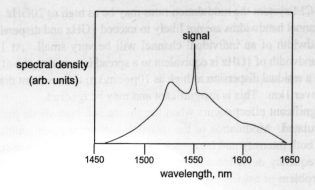

Figure 1.7   *Spectrum of light intensity at the output end of an erbium-doped fibre amplifier*

When the signal is detected there are several major mechanisms which can cause noise to appear in the signal band. Components of the ASE noise at different wavelengths can beat together and fall in band. This is known as spontaneous-spontaneous beat noise. Signal-spontaneous beat noise is a similar process occurring between the signal and the ASE noise and is usually the major noise component in practice. This process sets the theoretical minimum noise figure of an EDFA to 3dB and in practice noise figures very close to this have been measured.

1.4.2 Dispersion and distortion

Fibre dispersion does of course depend on the type of fibre used. Step index and graded index fibre can be used for short links but will introduce considerable amounts of intermodal dispersion. In most modern fibre distribution systems designed to operate over distances of more than a few km monomode fibre is used. Even in short-haul systems it is now becoming normal to use monomode fibre. The major disadvantage of monomode fibre, namely that of making permanent and demountable joints, has largely been solved and low-loss connectors (<0.5dB) are readily available. Of course monomode fibre exhibits a small amount of chromatic dispersion caused by a combination of material dispersion and waveguide dispersion. While this dispersion can in theory be reduced to zero by operating at 1.3μm, or at 1.55μm in dispersion-shifted fibre, in practice small amounts of residual dispersion do remain due to a variety of factors. For example, drift in the laser central frequency can cause the operating wavelength to deviate slightly from the zero dispersion wavelength. In practice residual dispersion of a few ps/nm/km is likely. The dispersion causes different optical wavelengths (frequencies) to travel at different velocities in the fibre.

In a typical SCM system the modulation rates may be as high as 20GHz. However individual channel bandwidths are not likely to exceed 1GHz and dispersion effects over the bandwidth of an individual channel will be very small. At 1.3μm, for example, a bandwidth of 1GHz is equivalent to a spread in wavelengths of 0.006nm. If we assume a residual dispersion as high as 10ps/nm/km the resultant dispersion is only 0.06ps over 1km. This is insignificant and may be ignored.

A more significant effect occurs when the source is a laser diode that has been directly modulated. Modulation of the injection current of a semiconductor laser diode causes both intensity and frequency modulation. In a DFB laser operating at 1.3μm the frequency deviation can be as high as several GHz/mA. In a digital system the problem of associated frequency modulation is well known. The laser frequency changes as the input pulse rises and falls on logic transitions. The phenomenon is known as chirp and causes the optical pulse to spread in the frequency domain, resulting in output pulses whose optical spectrum may be as wide as 100GHz. The different wavelengths travel at different velocities in the fibre, thus arriving at the receiver at different times causing the received pulse to be spread in the time domain.

In an analogue optical communication system the effects of chirp are rather different. Consider a semiconductor laser diode biased at the midpoint of its characteristic and driven with a sinusoidal signal at the subcarrier frequency. During one cycle of the injection current, as the optical power increases the output wavelength will decrease and as the optical power decreases the wavelength will increase. The arrival time of the optical carrier is thus dependent on the instantaneous optical power and this results in phase modulation of the carrier.

The effect can be explored by expanding the phase-modulated optical carrier as a series of Bessel functions and defining the second order harmonic distortion as:

$$2HD_{DIS} = \frac{J_1}{J_0} j \qquad\qquad (1.5)$$

where J_0 and J_1 are the zero and first order Bessel functions. The calculated dispersion-induced second order harmonic distortion for 400MHz/mA frequency chirping and 17ps/nm/km dispersion means that the composite second order distortion (CSO) value of -60dBc is only just met at 600MHz for a 30 km system [40].

When an optical amplifier is included in the system further distortions associated with laser chirp are produced. A typical fibre amplifier has a nonuniform gain profile with the same profile as the ASE curve of Figure 1.7 and the laser wavelength variations during modulation are transformed to signal variations at the output of the amplifier. It is straightforward to calculate the resultant second harmonic distortion

[41] and the results of this calculation indicate that CSO system requirements cannot be met unless the laser chirp is below 300MHz/mA and the laser frequency is controlled within the range 1.554 to 1.564 µm when an EDFA is used.

In a 1.3µm system the major causes of distortion are DFB laser distortion and (praseodymium-doped) fibre amplifier distortion. In a 1.55µm system the major causes of distortion are the chromatic dispersion and the (erbium-doped) fibre amplifier distortions. Fortunately both fibre amplifier induced distortion and chromatic dispersion induced distortion can be controlled with some simple electronic processing [41].

Of course, the distortion caused by laser chirp can be removed entirely if a CW laser and an external modulator are used, although the modulator will itself introduce new distortion as a result of its nonlinear characteristic. Alternatively, the effects of distortion can be reduced if FM rather than AM is chosen as the method for modulation of the subcarrier. The choice of system components, modulation methods and system architecture is determined by the particular application.

1.5 Detectors and receivers

Optical receivers consist of a photodetector, usually a photodiode, and an amplifier. The main requirements for the receiver are that it should convert the optical signal to an electrical signal efficiently, should be capable of operating with an appropriate bandwidth and should add as little noise as possible to the signal. Much attention has been focused on this final requirement since optical receivers traditionally work with very low level signals and careful design is required to produce low-noise wide-bandwidth receivers. The original studies by Personnick [42] and others have led to some standard approaches to optimising receiver performance for optical systems.

Most of the SCM systems proposed to date have used intensity modulation of the optical source with direct detection at the receiver. Of course, coherent optical communication techniques can be applied to subcarrier systems [5] providing the usual advantages of increased receiver sensitivity, increased link lengths and allowing larger numbers of users through dense wavelength division multiplexing. These systems do certainly offer many advantages but against this must be weighed the complexity inherent in coherent systems. Precise control of laser frequencies is required together with the use of narrow line (low phase noise) lasers. In contrast, in SCM systems employing intensity modulation/direct detection (IM/DD) methods the laser frequency does not need precise control and the linewidth of the laser has little effect on the detected signal. The selection of the optical source and the design of the receiver is considerably simplified. Furthermore, the sensitivity advantages provided by coherent techniques can be matched in IM/DD systems by including optical amplifiers. It is therefore only in specific situations, such as those where dense channel packing is required or where direct optical modulation, for example phase modulation, can be exploited, that coherent SCM systems are likely to be used.

For digital intensity-modulated direct-detection systems, avalanche photodiodes (APDs) are often used in order to improve receiver sensitivity. The gain of the APD is increased until the APD excess noise is the dominant noise process at the receiver. In SCM systems, however, the receivers are often narrow band with a consequent reduction in both thermal and amplifier noise. Signals from out of band subcarriers cause additional shot noise and the result is that narrow band SCM receivers can be operating in the shot-noise limited regime. This means that the use of an APD will not improve the noise performance of these receivers. In the case where broadband receivers are used, for example when receiver front ends capable of detecting all the transmitted subcarriers are needed, then the use of APDs may be advantageous.

PIN FET receivers are often used for digital IM/DD systems and can also be used in some SCM receivers where they provide good noise performance by minimising front-end capacitance. For narrow band receivers an inductor can be used to tune the front-end and reduce further the effects of capacitance [43]. However, if a broadband receiver is required then it is unlikely that a PIN FET approach will attain shot-noise limited performance across the full frequency range.

Another common approach to reducing receiver noise in digital optical systems is to use a transimpedance amplifier. This is a current to voltage amplifier and consists of a voltage amplifier with a feedback resistor between the output and the inverting input. It is probably the most common receiver structure in lower speed optical receivers but is difficult to use above 500MHz since phase shifts around the feedback loop increase with frequency and may cause oscillation. Since many SCM systems operate in the GHz regime transimpedance amplifiers are rarely used.

The most straightforward approach to a high frequency receiver is to use a photodiode with a bias resistor and a 50Ω microwave amplifier. These amplifiers are readily available with good noise figures. Although this approach does not provide the optimum receiver performance it is often the choice in practice for receivers in the microwave optical region.

In SCM receivers it is usually necessary to employ some form of down-conversion to extract the signal from the subcarrier. The photodiode and amplifier are followed by a mixer and local oscillator. However, down-converting PIN receivers which have the PIN detector connected directly to a mixer without an intervening stage of amplification can also be used. This provides a particularly simple receiver structure, especially if 50Ω mixers are used, but the conversion loss of the mixer will affect the noise performance and the impedance mismatch between the PIN and the mixer can also degrade the noise performance [44]. Typical conversion losses are in the region of 6-10dB. If shot-noise (or RIN) limited operation is achieved then the conversion loss need not have a severe effect on the noise performance since both the signal and the shot noise are attenuated by the same amount in the mixing process. With the 50Ω mixers, however, this is rarely the case and the conversion loss will usually result in a reduction in the SNR of the resultant IF frequency.

The problem of impedance matching can be attacked in a number of ways. A lumped element approach or the use of distributed elements in the form of microstrip can be employed. Impedance matching is usually only available for narrow-band receivers but can offer significant improvements in receiver sensitivity either for the down-converting receiver or for the conventional direct detection receiver. Improvements of up to 10dB have been achieved [45].

It is also possible to provide down-conversion directly at the photodetector by modulating either the PIN or APD bias or by using a device such as an FET or HBT and modulating the gate or base voltage to cause mixing [46]. Alternatively, the incoming optical signal can be used to frequency-lock an oscillator [47]. These techniques promise simple wideband SCM receiver architectures and may be of great importance where large numbers of low-cost receivers are required, for example in distribution networks.

In addition to their employment in optoelectronic mixing there has also recently been some interest in using three terminal devices, such as HBTs, as photodetectors in GHz optical receivers. Many of these devices operate well into the microwave range and are readily available. The area into which the optical signal is coupled is very small and there are significant coupling problems, but these are really no more severe than in high-speed photodiodes where the active region is also very small. These devices can provide noise improvements [48] because of their intrinsic gain and integrated structures based on this approach or on combining MSM diodes with custom receivers hold great promise.

An alternative and rather elegant approach to optical SCM receivers is the technique of optical prefiltering. Here an optical filter is placed before the photodiode and the resulting signal can be detected using baseband bandwidth photodiodes and receiver electronics. This approach has enormous potential for applications where very high frequency subcarriers are used or where wide bandwidth receiver electronics are to be avoided, for example in multi-user networks. In a demonstration experiment data modulated onto a 2GHz subcarrier was detected using a low bandwidth photodiode and amplifier: the amplifier bandwidth was a few kHz [49]. The technique can be used for both amplitude modulated and frequency modulated subcarriers. The principle is illustrated in Figure 1.8.

An optical signal which is intensity modulated with a microwave subcarrier will consist of an optical carrier and a pair of optical sidebands. If data is amplitude modulated onto the microwave subcarriers then data sidebands appear centred on the optical subcarrier. If a tunable optical filter is used to select a particular subcarrier and associated data sidebands, the photocurrent will consist of the original baseband information. This is a consequence of the optical filter blocking the energy from the other subcarriers with the result that the photodiode appears to be illuminated by a single intensity modulated signal. Hence the baseband information is detected directly by the prefiltered signal illuminating the detector. The photodetector and any subsequent processing electronics need only have a bandwidth consistent with the baseband signal. This contrasts strongly with conventional SCM demultiplexing where a photodetector and amplifier combination with a bandwidth up to the highest subcarrier frequency is required. The method can also deal with frequency modulated subcarriers [12]. In this case the linear part of the filter slope is aligned with the optical subcarrier and both demultiplexing and frequency discrimination can be achieved as shown in Figure 1.8.

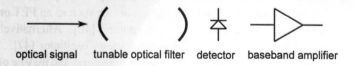

optical signal   tunable optical filter   detector   baseband amplifier

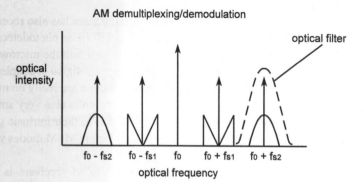

AM demultiplexing/demodulation

optical filter

optical
intensity

$f_0 - f_{s2}$ $f_0 - f_{s1}$ f_0 $f_0 + f_{s1}$ $f_0 + f_{s2}$

optical frequency

FM demultiplexing/discrimination

optical filter

optical
intensity

$f_0 - f_{s2}$ $f_0 - f_{s1}$ f_0 $f_0 + f_{s1}$ $f_0 + f_{s2}$

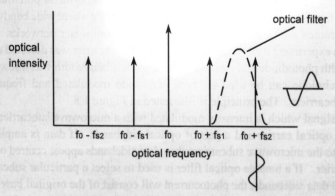

optical frequency

Figure 1.8   *The system configuration and resulting spectra for optical prefiltering of AM and FM modulated optical subcarriers*

Optical prefiltering is an interesting technique but does present some challenges. Firstly the filter must track any laser wavelength drift; fortunately this is fairly straightforward. A more significant issue is the effect of the shape of the filter passband on the baseband signal. An ideal optical filter for this kind of application would have a flat passband (for AM) or a linear slope (for FM). However, distortion introduced by the non-ideal filter characteristics can be equalised in the receiver by filtering with an inverse characteristic [50], although this will of course introduce a noise penalty.

1.6 Applications

1.6.1 Replacement of point-to-point RF/microwave links

An RF/microwave subcarrier can be considered for direct replacement of an RF/microwave carrier in an electrical (coaxial cable or waveguide) system. The use of optical fibre as the transmission medium brings a number of advantages. Probably the most significant advantage is the low incremental loss with link length; the main loss in a fibre system is typically in the electro-optic conversions at transmitter and receiver while the actual cable loss is about three orders of magnitude lower than for coaxial cable. This means that at microwave frequencies (and beyond) optical fibre-based systems can give lower insertion loss than coaxial cable systems for link lengths as short as a few tens of metres [23].

Another important feature of optical fibres is their light weight and small size. In systems where bulk is a concern, for example shipboard, avionic or space platforms, optical fibres become very attractive. This is especially true for complex systems such as array antennas where signals must be distributed between numerous components and subsystems. Their freedom from electromagnetic interference is another attractive feature in many such systems.

In considering the use of an optical fibre link, the RF/microwave engineer would be interested in a number of parameters apart from the insertion loss; these would include bandwidth, signal-to-noise ratio or a noise figure, linearity and dynamic range. Linearity and intermodulation distortion have been considered previously. Stephens and Joseph [23] and Cox *et al.* [51,52] have considered insertion loss and noise figure for general analogue optical links.

Non-dissipative impedance matching of both the source and detector improves link gain, as does an increase in their slope (or conversion) efficiencies. It should be noted, however, that impedance matching techniques will restrict system bandwidth [53]. Most analogue systems will be shot-noise limited; therefore, improving the detector has little or no effect on the noise figure, whereas impedance matching and/or increasing the slope efficiency of the source improves (i.e. decreases) the noise figure. These conclusions are true for both direct and external modulation [52]. A major difference between the two cases is when an increase in optical power is considered. For direct modulation, the link gain is not improved (the RF modulation on the laser is the same) and the increased shot noise means that the link noise figure is increased. Therefore, consideration of the noise figure means that directly modulated links should operate at an optical power level as low as possible consistent with the modulation conditions. For external modulation, although the shot noise still increases with optical power, the link gain increases quadratically due to the square-law nature of the photodetector; this means that the noise figure decreases (almost linearly) with increasing optical power [52].

1.6.2 Video distribution systems

Probably the major application area for optical fibre SCM systems up to the present has been in video distribution or CATV services. The transmission distances in such networks are relatively short, and the received optical power levels are such that shot noise or laser RIN are the dominant noise mechanisms. Linearity is an important concern even for moderate modulation depths as the systems typically involve the distribution of many channels; large numbers of intermodulation products fall at/near the subcarrier frequencies and the cumulative or *composite* effect of these products - composite second-order (CSO) and composite triple-beat (CTB) levels - must be taken into account. For example, with a 60- (equi-spaced) channel multiplex, over 1000 products may fall at many (predominantly mid-band) subcarriers while some tens of second-order products fall at some frequencies (predominantly at the ends of the multiplex) [19,54]. When fewer channels need to be transmitted using a multiplex occupying less than an octave of bandwidth is obviously advantageous as the effects of second-order products are avoided.

Most video distribution systems use vestigial sideband AM (AM-VSB) transmission for compatibility with ordinary television receivers. The noise and linearity requirements for these systems are stringent. Typical specifications for a trunk distribution network (of perhaps 40 to 80 channels) are CNR > 55dB, CTB < -65dBc and CSO < -55dBc [19]. The specifications for the local, fibre-to-the-home (FTTH) part of the network (which may transport 20 to 40 channels) are typically CNR > 48dB, CTB < -55dBc, CSO < -50dBc [19].

Alternative modulation formats can be used to relax the above requirements. For example, FM transmission reduces the required CNR to > 16.5dB at the expense of an approximate five-fold increase in bandwidth [55]. The bandwidth increase is significant for large numbers of channels. In the case of direct modulation the laser would have to be modulated at frequencies much closer to the laser relaxation-oscillation frequency, resulting in an increase in distortion. Nevertheless, for trunk systems especially, where noise may be a more significant limitation, FM offers distinctly enhanced performance. Olshansky *et al.* [55] have reported 60- and 120-

channel FM CATV systems with link distances of 18 km and 12 km, respectively. As satellite television broadcast systems also employ FM, a fibre network employing FM subcarrier transmission will allow direct entry of the satellite services into the local broadband loop. Mass production of components for the satellite television market and their resultant low cost makes the adoption of FM-SCM systems even more attractive.

Group modulation [56] has also been suggested to relax CATV requirements. Groups of individual subcarrier channels are modulated onto higher level subcarriers (with FM being used on these subcarriers). This can reduce intermodulation within groups and with a proper choice of frequency bands can reduce interference. However, it does have implications for the architecture of the network.

Digital video can also be distributed by using a digital modulation format such as FSK or PSK on the subcarriers. Uncompressed, each video channel requires a bandwidth of about 100 Mb/s; however, recent developments in the MPEG video compression standards should allow transmission at rates of less than 6 Mb/s. In the near term the simplicity of receiver equipment, its wide availability due to satellite broadcast services, and the very small difference in CNR requirements [55] suggest that FM video should be more attractive than FSK. However, the developments in video compression suggest that it will not be long before digital video becomes the preferred option [57]. Frequency modulation has also been considered for HDTV transmission over optical fibre; a 108 MHz bandwidth is required [58] compared to an uncompressed digital rate of 432 Mb/s. Again, compression techniques, which should bring the required rate down to about 20 Mb/s, should soon make digital modulation formats preferable [57].

1.6.3 Broadband switched services

In the longer term it is envisaged that broadband-ISDN (B-ISDN) services will be provided to each subscriber from digital connections using ATM (asynchronous transfer mode) -based interfaces [59]. The bit-rates involved (155 Mb/s and later perhaps beyond) are such that fibre-to-the-home (FTTH) topologies for the local loop become very attractive. Significant advantages can be gained by using SCM techniques to overlay the digital services network and the video distribution network (which was discussed previously in 1.6.2) [60,61]. However, in the near- to medium-term (over the next 5 years, say) it is unlikely that the services to utilise this bandwidth will fully mature; less expensive alternatives to the provision of fibre and optoelectronic components for each subscriber are therefore desirable. These alternative topologies are of the passive optical network (PON) and fibre-to-the-curb (FTTC) type, in which the initial costs of installation are reduced by sharing fibre and optoelectronic components amongst groups of subscribers.

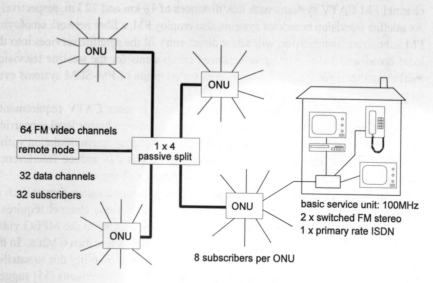

Figure 1.9   *Fibre-to-the-curb structure of an optical broadband services network using SCM, after [57]*

It is important that in designing for the near- and medium-term there is enough flexibility in the network for the longer term goals to be easily met. For example, Olshansky *et al.* [57] describe a broadband services network in which a remote node serves 32 subscribers. Eight subscribers are attached to each optical network unit (ONU); prior to this the transmission, including the 1:4 split, is optical. This fibre-to-the-curb type system is shown in Figure 1.9. Each subscriber is allocated 100 MHz bandwidth for the downstream switched services. Initially, this is seen as carrying two switched FM video channels (for video-on-demand type services) and a primary rate ISDN connection (1.5 Mb/s in North America, 2 Mb/s in Europe). In the optical network these channels are subcarrier multiplexed (at the remote node) between 1 and 4.2 GHz. The ONUs downconvert the required channels for their subscribers. Upstream primary rate ISDN connections are subcarrier multiplexed in a lower frequency range (10 - 106 MHz) at the ONUs. The possible evolution of the switched network services is then envisaged as shown in Figure 1.10. Initially, in (b), one of the FM video signals is replaced by a BPSK 20 Mb/s (compressed) HDTV signal. In (c), QPSK is used: two HDTV signals are transmitted in the bandwidth formerly occupied by an FM video signal, and a 45 Mb/s switched multi-megabit data service replaces the other FM video signal. Finally, in (d), a 155 Mb/s QPSK B-ISDN signal is transmitted in the whole of the 100 MHz subscriber channel.

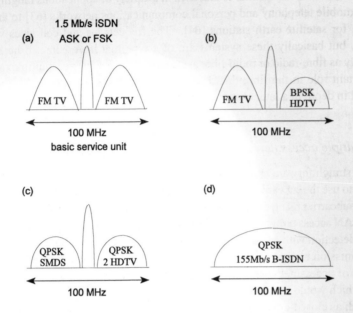

Figure 1.10 *Evolution of services using the 100MHz subscriber downstream allocations in a proposed optical broadband services network, after [57]*

There are also many possibilities for the simultaneous use of WDM in such systems. In the above system, the downstream channel employs 1.3μm transmitters and the upstream 1.55μm. It is also possible to use one wavelength to provide the switched/interactive service and the other to provide the broadcast distribution service (CATV) [57,62]. In systems where the same electrical frequencies are being used at both wavelengths, the crosstalk in components such as grating demultiplexers could be a serious limiting factor [62]. WDM techniques are considered in a later chapter.

1.6.4 Fibre-radio systems

In the previous subsection it was stated that costs will preclude the provision of fibre-to-the-home in the near term, and metallic cable final drops were most likely for the interim. An alternative to the metallic cable final drop would be a radio link. The network structure, with an optical fibre backbone connecting radio base station transceivers, is then very similar to that used in a variety of applications ranging from cordless/mobile telephony and personal communications networks [63] to antenna remoting for satellite earth stations [64]. The frequencies and cell sizes will be different, but basically these systems are of a type that have come to be known generically as fibre-radio or radio-fibre systems [7,63]. Subcarrier multiplexing has an important role in the fibre distribution network in these systems [7]; they are discussed in depth in a later chapter of this book.

1.6.5 Multiple access networks/LANs

The most straightforward application of SCM to local-area, multiple access networks would be to use the subcarrier as a data channel. The data frames could be modulated on to the subcarrier using digital modulation techniques such as ASK, FSK or PSK. Typical LAN access protocols would have to be used in some cases, for example with collision detection with some modification due to the differing nature of the medium and transmission technique. Several different subcarriers could be used to overlay a number of access networks on the same physical medium; however, the type of overlay which would be most attractive is that of a data network and other network types, such as closed-circuit TV distribution.

SCM can also be used as part of the multiple access technique itself. Such a network structure is shown schematically in Figure 1.11, where a star coupler is used to interconnect N users; users receive at a pre-allocated subcarrier frequency whereas they transmit at the subcarrier frequency corresponding to the required destination [3]. The architecture differs from typical LANs in that users have continuous access to the network but access to particular users needs to be controlled. A performance analysis of this type of network considering the power division and shot noise limitations suggests significant potential; for example, over 1000 users with data rates on each subcarrier of over 1.5 Mb/s. However, the real limitation to this type of structure is optical beat noise; the square-law photodetection process causes heterodyning of the laser frequencies and produces difference frequencies which may fall within the microwave bandwidth of the photodiode causing interference. With relatively few users, the laser wavelengths could be selected to minimise these effects but this will be impractical for any reasonably sized LAN community. Another alternative is to use low-coherence sources (such as light-emitting diodes, superluminescent diodes or self-pulsating lasers) but these will limit transmission distances due to fibre dispersion.

Techniques have also been proposed which eliminate optical beat noise by employing only a single laser in the multiple access network [65,66]. For example, in [65] the laser power is evenly distributed amongst the users in the star coupler used to interconnect them. Each user modulates their power at their own subcarrier frequency. These subcarrier transmissions will continuously circulate around the network but are severely attenuated for more than one pass and are therefore not a serious hindrance. However, the low level of each subcarrier compared to the total optical power level dictates that coherent SCM detection techniques are employed.

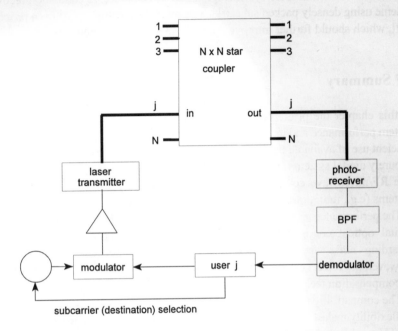

Figure 1.11   *An SCM multiple access LAN structure, after [3]*

Instead of using subcarriers to carry the data, it is also possible to use subcarriers to carry the control information (e.g., for access protocol implementation) while the data is transmitted in another format. For example, the use of baseband data packets and subcarrier multiplexed control channels has been both theoretically [67,68] and experimentally [69] investigated for WDMA networks. Transmitters use fixed-wavelength lasers while the receivers employ wavelength-tunable filters to select the required data channel. The transmitter will also transmit a subcarrier control "header" at the subcarrier frequency corresponding to the destination node it wants to reach. This header is detected prior to the wavelength filter, by tapping-off some of the incoming power and routing it to an SCM receiver; the header carries the information required to tune the wavelength filter to the correct wavelength. Two types of contention problems then exist: (1) the subcarrier header is sent to a node which is already listening to a transmission on another wavelength and therefore cannot re-tune its wavelength filter; (2) the subcarrier header is sent to a node at the same time as another node is transmitting a header to the same destination, causing a collision between the two headers. For optimum performance each node would be allocated its own subcarrier frequency at which to receive; however, this may be difficult in large networks and unnecessary depending on the traffic conditions. These problems have been analysed in [67]. Many aspects of the access protocols that can be used with these systems remain to be resolved. A contention recovery scheme using densely packed subcarrier acknowledgement tones has been proposed [70], which should further improve the performance of such networks.

1.7 Summary

In this chapter the principles behind SCM, the factors affecting component and system performance and its application potential have been discussed. SCM makes efficient use of available bandwidth and, even when techniques to reduce bandwidth in purely digital systems are considered, SCM remains less complex and expensive. The RF/microwave components required are often readily available from other systems (e.g., radio, radar, satellite) again reducing costs.

The performance requirements for SCM systems are different to those for typical digital optical communications systems. Generally, these are more stringent regarding noise and nonlinearities, especially for multichannel applications. However, the correct choice of components and system design, perhaps with the use of compensation techniques, does allow these requirements to be met.

The compatibility of SCM with many other types of RF/microwave systems and its flexibility makes it useful in a variety of applications, from antenna remoting and CATV to local-area networks. Probably the most significant feature of this flexibility, for the near-term application of SCM techniques, is its ability to allow an evolutionary development of existing networks into broadband-ISDN with low initial installed costs.

1.8 References

1 Tucker, R.S., Eisenstein, G., Korotky, S.K., Koren. U., Raybon, G., Veselka, J.J., Buhl, L.L., Kasper, B.L., and Alferness, R.C.: 'Optical time-division multiplexing and demultiplexing in a multigigabit/second fibre transmission system', *Electron. Lett.*, 1987, **23**, pp. 208-209

2 Wagner, S.S., and Lemberg, H.L.: 'Technology and system issues for a WDM-based fiber loop architecture', *J. Lightwave Technol.*, 1989, 7, pp. 1759-1768

3 Darcie, T.E.: 'Subcarrier multiplexing for multiple access lightwave networks', *J. Lightwave Technol.*, 1987, **5**, pp. 1103-1110

4 Darcie, T.E.: 'Subcarrier multipexing for lightwave networks and video distribution systems', *IEEE J. Selec. Areas Comms.*, 1990, **8**, pp. 1240-1248

5 Olshansky, R., Gross, R., and Schmidt, M.: 'Subcarrier multiplexed coherent lightwave systems for video distribution', *IEEE J. Selec. Areas Comms.*, 1990, **8**, pp. 1268-1275

6 Georges, J.B. and Lau, K.Y.: 'Broadband microwave fibre optic links with RF phase control for phased array antennas', *IEEE Photon. Technol. Lett.*, 1993, **5**, pp. 1344-1346

7 Wake, D., Westbrook, L.D., Walker, N.G., and Smith, I.C.: 'Microwave and millimeter-wave radio fibre', *BT Technol. J.*, 1993, **11**, pp. 76-88

8 Chiddix, J.A., Laor, H., Pangrac, D.M., Williamson, L.D., and Wolfe, R.W.: 'AM video on fibre in CATV systems: neeed and implementation', *IEEE J. Selec. Areas Comms.*, 1990, **8**, pp. 1229-1239

9 Bickers, L., Reeve, M.H., Rosher, P.A., Fenning, S.C., Cooper, A.J., Methley, S.G., and Hornung, S.: 'The analog local loop: a growing revolution in optical transmission', *J. Lightwave Technol.*, 1989, **7**, pp. 1819-1823

10 Agrawal, G.P., and Dutta, N.K.: 'Long-wavelength semiconductor lasers' (Van Nostrand Reinhold, New York, 1986)

11 Bjarklev, A.: 'Optical fiber amplifiers: design and system applications' (Artech House, Norwood, MA, 1993)

12 Way, W.I., Lo, Y.H., Lee, T.P., and Lin, C.: 'Direct detection of closely spaced optical FM-FDM Gb/s Microwave PSK signals', *IEEE Photon. Technol. Lett.*, 1991, **3**, pp. 176-178

13 Gross, R., Olshansky, R., and Hill, P.: '20 channel coherent FSK system using subcarrier multiplexing', *IEEE Photon. Technol. Lett.*, 1989, **1**, pp. 224-226

14 Nazarathy, M., Berger, J., Ley, A.J., Levi, I.M., and Kagan, Y.: 'Progress in externally modulated AM CATV transmission systems', *J. Lightwave Technol.*, 1993, **11**, pp. 82-105

15 Saleh, A.A.M.: 'Fundamental limit on number of channels in subcarrier-multiplexed lightwave CATV system', *Electron. Lett.*, 1989, **25**, pp. 776-777

16 Alameh, K., and Minasian, R.A.: 'Ultimate limits of subcarrier multiplexed lightwave transmission', *Electron. Lett.*, 1991, **27**, pp. 1260-1262

17 Frigo, N.J., Phillips, M.R., and Bodeep, G.E.: 'Clipping distortion in lightwave CATV systems: models, simulations, and measurements', *J. Lightwave Technol.*, 1993, **11**, pp. 138-146

18 Morthier, G.: 'Influence of the carrier density dependence of the absorption on the harmonic distortion in semiconductor lasers', *J. Lightwave Technol.*, 1993, **11**, pp. 16-19

19 Darcie, T.E., and Bodeep, G.E.: 'Lightwave subcarrier CATV transmission systems', *IEEE Trans.*, 1990, **MTT-38**, pp. 524-533

20 Morton, P.A., Tanbun-ek, T., Logan, R.A., Sciortino, P.F., Sergent, A.M., and Wecht, K.W.: 'Superfast 1.55µm DFB lasers', *Electron. Lett.*, 1993, **29**, pp. 1429-1430

21 Oberg, M., Kjebon, O., Lourdudoss, S., Nilsson, S., Backbom, L., Streubel, K., and Wallin, J.: 'Increased modulation bandwidth up to 20 GHz of a detuned-loaded DBR laser', *IEEE Photon. Technol. Lett.*, 1994, **6**, pp. 161-163

22 Way, W.I.: 'Subcarrier multiplexed lightwave system design considerations for subscriber loop applications', *J. Lightwave Technol.*, 1989, **7**, pp. 1806-1818

23 Stephens, W.E., and Joseph, T.R.: 'System characteristics of direct modulated and externally modulated RF fiber-optic links', *J. Lightwave Technol.*, 1987, **5**, pp. 380-387

24 Johnson, L.M., and Rousell, H.V.: 'Reduction of intermodulation distortion in interferometric optical modulators', *Optics Lett.*, 1988, **13**, pp. 928-930

25 Wang-Boulic, Y.: 'A linearized optical modulator for reducing third-order intermodulation distortion', *J. Lightwave Technol.*, 1992, **10**, pp. 1066-1070

26 Korotky, S.K., and de Ridder, R.M.: 'Dual parallel modulation schemes for low-distortion analog optical transmission', *IEEE J. Selec. Areas Comms.*, 1990, **8**, pp. 1377-1381

27 Farwell, M.L., Lin Z.-Q., Wooten, E., and Chang, W.S.C.: 'An electrooptic intensity modulator with improved linearity', *IEEE Photon. Technol. Lett.*, 1991, **3**, pp. 792-795

28 Pan, J.J., and Garafalo, D.A.: 'Microwave high dynamic range EO modulators', *Proc. Int. Soc. Optical Eng.*, **1371**, 1990, pp. 21-35

29 Gomes, N.J., Dye, S.P., and Davies, P.A.: 'Electronic compensation of optical intensity modulator nonlinearity: a comparison of techniques', EFOC/LAN'92, Paris, 1992, Digest, pp. 436-441

30 Trisno, Y., Chen, L.K., and Huber, D.: 'A linearized external modulator for analog applications', *Proc. Int. Soc. Optical Eng.*, **1371**, 1990, pp. 8-12

31 de Ridder, R.M., and Korotky, S.K.: 'Feedforward compensation of integrated optic modulator distortion', OFC'90, 1990, WH5

32 Ho, K.-P., and Kahn, J.M.: 'Equalization technique to reduce clipping-induced nonlinear distortion in subcarrier-multiplexed lightwave systems', *IEEE Photon. Technol. Lett.*, 1993, **5**, pp. 1100-1103

33 Fock, L.S., and Tucker, R.S.: 'Simultaneous reduction of intensity noise and distortion in semiconductor lasers by feedforward compensation', *Electron. Lett.*, 1991, **27**, pp. 1297-1299

34 Dolfi, D., Nazarathy, M., and Jungerman, R.: '40 GHz electro-optic modulator with 7.5 V drive voltage', *Electron. Lett.*, 1988, **24**, pp. 528-529

35 Lima, C.R., and Davies, P.A.: 'Noise performance of microwave signals generated by a gain-switched semiconductor laser', IEE Colloq. Microwave Optoelectronics, 1994, Digest No. 1994/022, pp. 2/1-5

36 Bates, R.J.S., and Walker, S.D.: 'Optical networking and signal-processing applications of 790 nm self-pulsating laser diodes', *IEE Proc.-J: Optoelectronics*, 1992, **139**, pp. 263-271

37 O'Reilly, J.J., Lane, P.M., Heidermann, R., and Hofstetter, R.: 'Optical generation of very narrow linewidth millimetre wave signals', *Electron. Lett.*, 1992, **28**, pp. 2309-2310

38 Ni, D.C., Fetterman, H.R., and Chew, W.: 'Millimeter-wave generation and characterization of a GaAs FET by optical mixing', *IEEE Trans.*, 1990, **MTT-38**, pp. 608-614

39 Goldberg, L., Yurek, A., Taylor, H.F., and Weller, J.F.: '35 GHz microwave signal generation with an injection locked laser diode', *Electron. Lett.*, 1985, **21**, pp. 714-715

40 Kuo, C.Y.: 'Fundamental nonlinear distortions in analog links with fibre amplifiers', *J. Lightwave Technol.*, 1993, **11**, pp. 7-15

41 Kuo, C.Y. and Bergmann, E.E.: 'Erbium doped fibre amplifier second-order distortion in analog links and electronic compensation', *IEEE Photon. Technol. Lett.*, 1991, **3**, p. 829

42 Personick, S.D.: 'Receiver design for digital fibre optic communication systems, I and II', *Bell System Tech. J.*, 1973, **52**, pp. 843-86

43 Moreira, P.M.R.S., Darwazeh, I., and O'Reilly, J.J.: 'Design and optimisation of a fully integrated GaAs tuned receiver preamplifier MMIC for optical SCM applications', *IEE Proc.-J: Optoelectronics*, 1993, **140**, pp. 411-415

44 Urey, Z., Gomes, N.J., Davies, P.A., and Urey, H.: 'Optoelectronic mixing at 1300nm: performance comparison of an InGaAs/InP heterojunction FET and an InGaAsPIN photodiode', Bilcon 1992, Ankara, Turkey, 27-30 July 1992

45 Ackerman, E., Kasemet, D., Wanuga, S., Hogue, D., Komiak, J.: 'A high gain directly modulated L-band microwave optical link', 1990 IEEE MTT-S Int. Microwave Symposium Digest, pp. 153-155

46 Urey, Z. and Davies, P.A. : 'Optoelectronic downconversion of intensity modulated optical signals using GaAs MESFETs', EFOC/LAN'92, Paris, 1992, Digest, pp. 81-85

47 Sommer, D., Gomes, N.J., and Wake, D.: 'Optical injection locking of microstrip MESFET oscillator using heterojunction phototransistors', *Electron. Lett.*, 1994, **30**, pp. 1097-1098

48 Urey, Z., Wu, D., Gomes, N.J., and Davies, P.A.: 'Noise performance of a GaAs MESFET as an optical detector and as an optoelectronic mixer in analogue optical links', *Electron. Lett.*, 1993, **29**, pp. 147-149

49 Greenhalgh, P.A., Abel, R.D., and Davies, P.A.: 'Optical prefiltering in subcarrier systems', *Electron. lett.*, 1992, **28**, pp. 1850-1852

50 Greenhalgh, P.A., Foord, A.P. and Davies, P.A.: 'Signal distortion in optically prefiltered subcarrier systems', IEEE/LEOS 1994 Summer Topical Meeting on "Optical Networks and Their Enabling Technologies", July 11-13, 1994, Lake Tahoe, NV, USA

51 Cox, C.H., Betts, G.E. and Johnson, L.M.: 'An analytic and experimental comparison of direct and external modulation in analog fiber-optic links', *IEEE Trans.*, 1990, **MTT-38**, pp. 501-509

52 Cox, C.H.: 'Gain and noise figure in analogue fibre-optic links', *IEE Proc.-J: Optoelectronics*, 1992, **139**, pp. 238-242

53 Goldsmith, C.L., and Kanack, B.: 'Broad-band reactive matching of high-speed directly modulated laser diodes', *IEEE Microwave Guided Wave Lett.*, 1993, **3**, pp. 336-338

54 Lipson, J., Upadhyayula, L.C., Huang, S.-Y., Roxlo, C.B., Flynn, E.J., Nitzsche, P.M., McGrath, C.J., Fenderson, G.L., and Schaefer, M.S.: 'High-fidelity lightwave transmission of multiple AM-VSB NTSC signals', *IEEE Trans.*, 1990, **MTT-38**, pp. 483-493

55 Olshansky, R., Lanzisera, V.A., and Hill, P.M.: 'Subcarrier multiplexed lightwave systems for broad-band distribution', *J. Lightwave Technol.*, 1989, **7**, pp. 1329-1341

56 Berceli, T., Frigyes, I., Gottwald, P., Herczfeld, P.R., and Mernyei, F.: 'Performance improvements in fiber optic links for multi-carrier TV transmission', 1991 IEEE MTT-S Int. Microwave Symposium Digest, pp. 307-310

57 Olshansky, R., Lanzisera, V.A., Su, S.-F., Gross, R., Forucci, A.M., and Oakes, A.H.: 'Subcarrier multiplexed broad-band service network: a flexible platform for broad-band subscriber services', *J. Lightwave Technol.*, 1993, **11**, pp. 60-69

58 Senior, J.M., Lambert, D.T., and Faulkner, D.W.: 'Analogue intensity modulation schemes for optical fibre HDTV transmision', *IEE Proc.-J: Optoelectronics,* 1993, **140**, pp. 417-424

59 Chao, H.J., Shtirmer, G., and Smoot, L.S.: 'H-Bus: an experimental ATM-based optical premises network", *J. Lightwave Technol.*, 1989, **7**, pp. 1859-1867

60 Lo, C.N., Tohme, H.E., and Wolff, R.S.: 'A hybrid architecture for analog video broadcast and B-ISDN services in customer premises networks', *IEEE J. Selec. Areas Comms.*, 1990, **8**, pp. 1327-1339

61 Cheung, K.-W.: 'An evolutionary transport structure for local loop distribution using RF subcarriers', *IEEE J. Selec. Areas Comms.*, 1990, **8**, pp. 1340-1350

62 van Heijningen, P.H., Muys, W., van der Plaats, J.C., and Willems, F.W.: 'Crosstalk in a fibre access demonstrator carrying television and interactive digital services', *Electron. Comm. Eng. J.*, February 1994, **6**, pp. 49-55

63 Cooper, A.J.: '"Fibre/radio" for the provision of cordless/mobile telephony services in the access network', *Electron. Lett.*, 1990, **26**, pp. 2054-2056

64 Way, W., Wolff, R.S., and Krain, M.J.: 'A 1.3μm 35km fiber optic microwave transmission system for satellite earth stations', *J. Lightwave Technol.*, 1987, **5**, pp. 1325-1332

65 Jiang, Q., and Kavehrad, M.: 'An optical multiaccess star network using subcarrier multiplexing', *IEEE Photon. Technol. Lett.*, 1992, 4, pp. 1163-1165

66 Domon, W., Shibutni, M., and Emura, K.: 'SCM optical multiple-access network with cascaded optical modulators', *IEEE Photon. Technol. Lett.*, 1993, **5**, pp. 1107-1109

67 Su, S.F., and Olshansky, R.: 'Performance of multiple access WDM networks with subcarrier multiplexed control channels', *J. Lightwave Technol.*, 1993, **11**, pp. 1028-1033

68 Poggiolini, P., and Benedetto,S.: 'Performance analysis of multiple subcarrier encoding of packet headers in quasi-all-optical WDM networks', *IEEE Photon. Technol. Lett.*, 1994, **6**, pp. 112-114

69 Su, S.F., Bugos, A.R., Lanzisera, V., and Olshansky, R.: 'Demonstration of a multiple-access WDM network with subcarrier-multiplexed control channels', *IEEE Photon. Technol. Lett.*, 1994, **6**, pp. 461-463

70 Su, S.F., and Olshansky, R.: 'Use of subcarrier acknowledgement tones for contention recovery in WDMA networks', *Electron. Lett.*, 1993, **29**, pp. 1099-1100

Chapter 2

Coherent techniques in analogue signal transmission

A. J. Seeds

2.1 Introduction

Most optical fibre analogue transmission systems use intensity modulation of the optical source with direct detection of the modulated optical signal in a depletion photo-detector. In this chapter, coherent techniques are introduced and their advantages and disadvantages relative to intensity modulation/direct detection (IMDD) systems considered. Coherent techniques will be defined broadly to include all transmission methods where the precise frequency of the optical signal is important.

Section 2.2 will introduce analogue coherent transmission systems and contrast their performance with that of IMDD systems. In Section 2.3 optical source requirements for coherent analogue transmission will be discussed and a novel technique for generating optical frequency modulation (OFM) described. Section 2.4 will discuss coherent analogue transmission techniques. Section 2.5 will present results for an experimental coherent analogue transmission system using OFM. Finally, Section 2.6 will outline future prospects for coherent analogue transmission systems.

2.2 Coherent analogue transmission systems

Figure 2.1 contrasts IMDD and coherent optical transmission systems. In the IMDD system, shown in Figure 2.1(a), the optical source intensity is either directly modulated by the input analogue signal or passes through an external intensity modulator. The resulting intensity modulated signal is then transmitted along the optical fibre to the photodiode where the modulation is returned to the electrical domain.

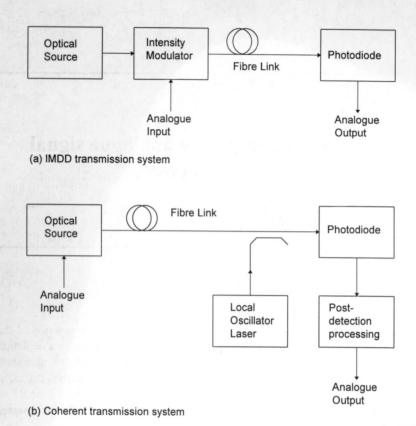

(a) IMDD transmission system

(b) Coherent transmission system

Figure 2.1 *IMDD and coherent analogue fibre transmission systems*

In the coherent system, Figure 2.1(b), the optical source is modulated in intensity, frequency or phase by the input analogue signal, either directly or by passage through an external modulator. The modulated signal passes through the optical fibre to the receiver, where it is combined with the output from a local oscillator (LO) laser. The combined signal illuminates the photodiode to produce an electrical signal centered on the difference frequency between the unmodulated optical source and the LO laser. This signal is then further processed to recover the analogue input signal.

2.2.1 Transmission system model - direct detection

Consider first the IMDD system shown in Figure 2.1(a). Let the analogue signal to be transmitted be represented by $m\{t\}$. The optical power at the output of the intensity modulator is :

$$P_o = P_u(1 + sm\{t\})$$ (2.1)

where P_u is the mean optical power and s is the modulation sensitivity ($s\ m\{t\} > -1$). The mean squared signal current at the detector output is:

$$I_s^2 = (RG_l P_u s)^2 \overline{m^2\{t\}}$$ (2.2)

where R is the photodiode responsivity and G_l the fibre path gain ($G_l < 1$ unless optical amplifiers are used). Noise at the detector output arises from several sources, including:

(i) Thermal noise current generated in the photodiode load, with mean square value:

$$I_{nt}^2 = \frac{4kTB}{R_L}$$ (2.3)

where k is Boltzmann's constant, T the absolute temperature, B the bandwidth and R_L the load resistance value;

(ii) Shot noise generated in the photodiode:

$$I_{ns}^2 = 2e(\overline{i_d} + i_{dk})B$$ (2.4)

where e is the electronic charge, $\overline{i_d}$ is the mean optically generated current in the photodiode and i_{dk} is the photodiode dark current;

(iii) Optical source relative intensity noise:

$$I_{nRIN}^2 = \overline{i_d}^2\, RIN\, B$$ (2.5)

where RIN is the source relative intensity noise value;

(iv) Noise generated by any optical amplifiers used, I_{na}^2.

Assuming these noise sources to be uncorrelated, the signal-to-noise ratio at the detector output can be written:

$$SNR = \frac{\left(RG_l P_u s\right)^2 \overline{m^2\{t\}}}{\left(\dfrac{4kT}{R_L} + 2e\left(RG_l P_u + i_{dk}\right) + \left(RG_l P_u\right)^2 RIN\right)B + I_{na}^2} \tag{2.6}$$

It can be seen that the thermal noise contribution is independent of unmodulated optical power, P_u, as is the optical amplifier noise contribution if non-linear effects can be neglected. Thus the signal-to-noise ratio can be improved by increasing the unmodulated optical power until the source relative intensity noise limit is reached, giving:

$$SNR = \frac{s^2 \overline{m^2\{t\}}}{RIN\,B} \tag{2.7}$$

For optical powers below the RIN limit, shot-noise-limited reception can be achieved if the thermal and optical amplifier contributions are sufficiently small, resulting in:

$$SNR = \frac{RG_l P_u s^2 \overline{m^2\{t\}}}{2eB} \tag{2.8}$$

where the photodiode dark current has been assumed negligible relative to the photocurrent. For low received optical powers, thermal and optical amplifier contributions are dominant.

2.2.2 Transmission system model - coherent detection

Consider now the coherent system, Figure 2.1(b), and assume that polarisation control techniques are used so that the signal and local oscillator electric fields incident on the photodiode have the same polarisation. The signal electric field is defined by:

$$E_s = \hat{E}_s \cos\left(\omega_s t + \phi_s\right) \tag{2.9}$$

where ω_s is the signal frequency and ϕ_s the signal phase, and the local oscillator field by:

$$E_{LO} = \hat{E}_{LO} \cos\left(\omega_{LO} + \phi_{LO}\right) \tag{2.10}$$

with ω_{LO} the local oscillator frequency and ϕ_{LO} the local oscillator phase. Defining the intermediate frequency (IF), ω_I, by $\omega_I = \omega_{LO} - \omega_s$ the analytic signal incident on the photodiode is:

$$V_{in} = \left(\hat{E}_s \exp j\phi_s + \hat{E}_{LO} \exp j(\omega_I + \phi_{LO}) \right) \exp j\omega_s t \qquad (2.11)$$

For $\omega_I \ll \omega_s$ the output current from the photodiode is proportional to $V_{in} V_{in}^*$ so that:

$$i \propto \hat{E}_s^2 + \hat{E}_{LO}^2 + 2\hat{E}_s \hat{E}_{LO} \cos\left(\omega_I t + \phi_{LO} - \phi_s\right) \qquad (2.12)$$

It is convenient to re-write Equation 2.12 in terms of optical power since that is a directly measurable quantity. The relationships:

$$\hat{E}_s^2 = \frac{2Z_o P_o G_I}{A} \qquad (2.13)$$

and

$$\hat{E}_{LO}^2 = \frac{2Z_o P_{LO}}{A} \qquad (2.14)$$

where Z_o is the impedance of the medium where the power is measured, P_o is the source output power and A is the photodiode area, give the photodiode current as:

$$i = R\left(P_o G_I + P_{LO} + 2\sqrt{P_o G_I P_{LO}} \cos\left(\omega_I t + \phi_{LO} - \phi_s\right) \right) \qquad (2.15)$$

The first two terms represent direct detection of the signal and local oscillator respectively. The third term is of more interest. First, its magnitude is proportional to the square root of the local oscillator power, thus the detected signal can be made larger simply by increasing the local oscillator power. Second, the detected signal is proportional to the square root of the source output power; thus linear modulation of the source electric field will yield linear modulation of the detected photo-current at the intermediate frequency. Alternatively, linear modulation of the source intensity will yield linear modulation of the output of a square-law detector fed with the photo-detected intermediate frequency (IF) signal. Third, the term is at the IF, ω_I, so that modulation of the source frequency, ω_s, leads directly to modulation of the IF, which can be recovered using a suitable discriminator. Fourth, the term contains the signal phase, ϕ_s, and the local oscillator phase, ϕ_{LO}, so that phase modulation of the source leads directly to phase modulation of the IF output. Thus, coherent systems can use intensity, frequency or phase modulation while direct detection systems are limited to intensity modulation. When $\omega_I = 0$ the coherent system is said to be homodyne but when $\omega_I \neq 0$ it is said to be heterodyne.

The sources of noise in a coherent system are similar to those in a direct detection system (see Section 2.2.1) giving an IF carrier-to-noise ratio after photo-detection of:

$$CNR = \frac{2P_oG_lP_{LO}R^2}{\left(\dfrac{4kT}{R_L} + 2e\left(RP_{LO} + i_{dk}\right) + \left(RP_{LO}\right)^2 RIN\right)B + I_{na}^2} \qquad (2.16)$$

where it is assumed that $P_{LO} \gg P_oG_l$ and RIN is the relative intensity noise of the local oscillator laser. It is normal practice in coherent receivers to use a balanced detection scheme to cancel LO laser RIN [1], so that by increasing the LO laser power shot-noise-limited reception is obtained, giving:

$$CNR = \frac{P_oG_lR}{eB} \qquad (2.17)$$

2.2.3 Comparison between coherent and direct detection systems

Coherent transmission systems offer three main advantages over systems using direct detection. These may be summarised as:

(i)   Shot-noise-limited reception can be achieved even at low received signal powers, simply by increasing the local oscillator power;

(ii)   Intensity, frequency or phase modulation modes can be used whereas direct detection systems are limited to intensity modulation;

(iii)   The excellent frequency selectivity that can be achieved using electrical post-photodetector filters is translated into the optical domain by the coherent detection technique, enabling the realisation of dense wavelength division multiplex schemes for multi-channel transmission or channel selection schemes.

Reviewing these advantages in turn; the first is of reduced importance for systems operating at a wavelength of 1550 nm now that effective optical amplifiers are available (see Chapter 11). However, there is interest in systems operating at 1300 nm wavelength in order to take advantage of the silica fibre dispersion minimum and the low noise and high output power of semiconductor laser pumped Nd-YAG lasers. There is also interest in systems operating at 850 nm wavelength for compatibility with GaAs microwave monolithic integrated circuit (MMIC) technology. Effective optical fibre amplifiers are not available for either of these wavelengths. The alternative strategy for shot-noise-limited IMDD systems of increasing the source power P_u is limited by the onset of stimulated Brillouin scattering (SBS) and other non-linear effects in optical fibre [2]. For a transmission distance of 30 km, the SBS threshold is as low as a few milliwatts, so that the high source power strategy is limited to system lengths of a few km or less. Thus coherent transmission remains of interest for medium and long distance systems operating at wavelengths other than 1550 nm.

The second advantage enables SBS to be reduced by broadening the optical signal bandwidth beyond the SBS linewidth (~28 MHz at 1300 nm wavelength) using frequency or phase modulation. Use of frequency or phase modulation also enables a trade-off to be made between optical signal bandwidth and received signal-to-noise ratio, an example of which is given in Section 2.5.

The importance of the third advantage depends upon whether the ability to switch between many signals carried on the same fibre is required. An example of such a requirement would be a distributed receive antenna remoting application.

There are three main disadvantages of coherent transmission systems relative to those using direct detection :

(i)   The frequencies of the local oscillator laser and signal must be controlled to differ by the required IF, whereas in the direct detection system it is only necessary that the source wavelength be suitable for the photodiode used;

(ii)   The linewidths of source and local oscillator lasers must be suitable for the modulation mode used, whereas in direct detection systems the required source linewidth is mainly determined by the optical fibre dispersion penalty;

(iii) The polarisation state of the local oscillator and signal must be matched at the photodiode.

The requirement for source frequency control is an exacting one. An operating wavelength of 1550 nm corresponds to a frequency of 194 THz, so that to maintain an heterodyne signal within 10% of band centre in an IF bandwidth of 2 GHz, control to within 1 part in 10^6 is required. For semiconductor lasers with typical temperature tuning sensitivities of 30 GHz K^{-1} and current tuning sensitivities of 3 GHz mA^{-1} this requires temperature control to within 7 mK and current control to within 7 μA.

The source linewidth requirement will be considered in more detail in Section 2.4. Advanced semiconductor lasers can offer linewidths in the kHz region coupled with wavelength tuning ranges in excess of 10 nm [3], although the commercial availability of such lasers is currently limited. Homodyne systems require the local oscillator frequency to be phase-locked to that of the received signal in an optical phase-lock loop (OPLL). Realising such loops with other than narrow linewidth lasers presents formidable challenges [4].

Polarisation matching can be achieved by active polarisation control of the local oscillator signal for maximum detected signal output [5] or using polarisation diversity reception [6].

Whilst the disadvantages of coherent transmission systems can all be overcome, the penalty is a significant increase in system complexity relative to direct detection systems. Whether the coherent system approach is used in a particular application therefore depends upon whether the performance advantages are sufficient to justify the increase in complexity.

2.3 Optical sources for coherent analogue transmission

Key performance parameters for sources used in coherent analogue transmission systems can be summarised as:

(i)   Output wavelength

(ii)   Output power

(iii)   Linewidth and intensity noise

(iv)   Tunability

(v)   Frequency modulation response

Fibre lasers, crystal and waveguide lasers and semiconductor lasers constitute the main families of optical sources suitable for coherent analogue transmission. Of these, only semiconductor lasers offer a direct modulation capability. The capabilities and limitations of these technologies are discussed below.

2.3.1 Fibre lasers

Optically pumped fibre doped with an appropriate lasing ion has made a key contribution to optical communication through the optical fibre amplifier and with appropriate feedback can be used to realise optical fibre sources.

The required length of doped fibre depends on cavity losses, pump power and lasing ion doping concentration. The latter is limited by clustering, so that practical erbium-doped fibre lasers for operation at 1530 nm wavelength typically require doped-fibre lengths of 1 to 5 m, giving laser mode separations of 100 MHz to 20 MHz. Obtaining stable single mode operation is therefore difficult. Unidirectional ring configurations, which eliminate spatial hole burning, can offer linewidths in the kHz region [7], but active stabilisation of the fibre length and polarisation state is required to eliminate mode hopping.

2.3.2 Crystal and waveguide lasers

Diode-pumped crystal lasers are attractive sources of high power (> 100 mW), narrow linewidth (< 1 kHz), low *RIN* (< -170 dBc Hz^{-1}) optical signals [8]. They are finding application as sources in wide dynamic range direct detection analogue optical transmission systems for cable TV and related applications. For general application, they suffer from limited tunability (< 60 GHz) and high optical complexity. However, the relatively narrow linewidth and good frequency stability have made them the main vehicle for early work on optical phase lock loops [9].

Improved optical confinement and potential for integration with other optical and electro-optic components make rare-earth-doped waveguide lasers of interest. The two main fabrication technologies are silica on silicon and titanium in-diffused lithium niobate. To realise lasers with reasonably small chip area it is necessary to achieve doping densities much higher than for fibre lasers. This is especially challenging for erbium-doped guides. Kitagawa *et al.* have achieved an output power of 1-2 mW from a 45 mm long erbium-doped silica on silicon laser pumped at 980 nm wavelength [10]. In-diffusion has been used to realise an erbium-doped laser in lithium niobate. A waveguide of length 10.5 mm gave a threshold of 8 mW for pumping at 1477 nm wavelength [11]. Non-semiconductor waveguide lasers require an optical pump source, which is generally desired to be a diode laser. Hence a fully integrated source technology is not yet possible.

2.3.3 Semiconductor lasers

Semiconductor lasers are used as sources in most current optical communication systems. They offer the advantages of electrical pumping and direct modulation capability. Key performance aspects requiring improvement for coherent analogue optical transmission systems are :

(i)   Narrow linewidth operation

(ii)   Tunability for wavelength matching

(iii)   Uniform FM frequency response for phase-locked applications.

Careful design of the laser structure to minimise spatial hole burning effects has led to the realisation of multiple quantum well (MQW) distributed feedback (DFB) lasers having linewidths below 100 kHz [3].

Renewed interest in wavelength division multiplexed (WDM) systems, both for increased channel capacity and for their use in all-optical routeing, has placed emphasis on the need for widely tunable laser sources. A number of approaches are possible. Multi-section DFB lasers have been realised with tuning ranges of 10 nm, while maintaining linewidths less than 20 MHz [12]. Wider tuning ranges have been achieved in vertically coupled structures [13], culminating in the 55 nm tuning range reported by Kim *et al.* [14], although the linewidth obtained was not reported.

A novel approach, yielding discontinuous tuning over a wide range with simple control, is the Y junction laser [15], where vernier effects have been used to increase the tuning range to 38 nm.

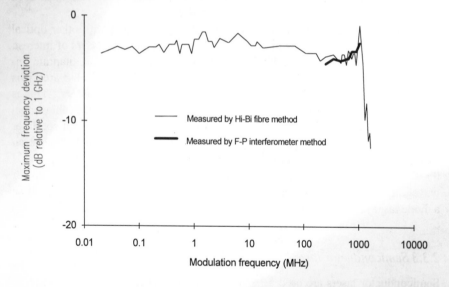

Figure 2.2 *FM frequency response of reverse-bias quantum-well-tuned external-cavity semiconductor laser [17]*

A major difficulty when semiconductor lasers are used as optical FM sources and current controlled oscillators (CCO) in phase-locked applications is that they exhibit a highly non-uniform FM response as a consequence of the interaction between thermal and carrier density effects. The most common approach to achieving uniform FM response has been to use a multi-section DFB laser with the section currents adjusted for best uniformity of response. Using this approach Ogita *et al.* [16] have achieved a -3 dB bandwidth of 15 GHz, although neither the peak frequency deviation nor the residual intensity modulation is stated. An alternative approach, capable of an intrinsically flat FM response, is to use the refractive index change resulting from the quantum-confined Stark effect (QCSE) in a reverse-biased multiple quantum well (MQW) section. This has already been demonstrated in external cavity laser systems [17]. Figure 2.2 shows the measured frequency response for such a laser over the frequency range 20 kHz to 1.5 GHz and it is seen that a response uniformity of ±1.6 dB over the frequency range 20 kHz to 1.3 GHz is obtained.

The rapid roll-off in response above 1.3 GHz is due to a resonance between the tuning element capacitance and package parasitics, however, with optimised design operation to millimetre-wave frequencies should be possible.

Reviewing the state of semiconductor laser research, it is clear that widely tunable sources with linewidths small enough for use in coherent analogue systems are realisable, although not yet widely available commercially.

2.4   Coherent analogue modulation formats

As discussed in Section 2.2, coherent analogue transmission systems offer the choice of intensity, amplitude, frequency or phase modulation. To date demonstrator systems have been reported for intensity, frequency and phase modulation. In this section the relative merits of the various modulation formats are discussed.

2.4.1 Intensity modulation

Reference to Equation 2.15 shows that if the source is intensity modulated and heterodyne detection followed by envelope detection is used, the output is proportional to $\sqrt{P_o}$ so that a square-law detector must be used to provide overall system linearity. Fong *et al.* have demonstrated this approach using semiconductor laser sources [18]. The system has the attraction that the main linewidth penalty arises from heterodyned source power falling outside the IF bandwidth. Thus by using a wide IF filter bandwidth and a post-detection filter the system can be made relatively insensitive to source linewidth and drift in source and local oscillator laser frequencies, allowing commercially available semiconductor lasers with linewidths of greater than 10 MHz to be used.

The system offers the advantage of improved received sensitivity through coherent detection and thus is useful at wavelengths where good optical pre-amplifiers are not available. System linearity is limited by the linearity of the source intensity modulation characteristic and by the accuracy of the square law detector characteristic. It therefore can be no better than for an IMDD system.

2.4.2 Amplitude modulation

In an amplitude modulation system the peak electric field produced by the source is linearly proportional to the modulating signal. Thus the source power is proportional to the square of the modulating signal. From Equation 2.15 it can be seen that heterodyne detection followed by enveloped detection would give an output linearly proportional to the modulating signal, removing the need for a square law detector. Unfortunately, directly modulated semiconductor lasers have a near-linear intensity modulation characteristic, whilst interferometric external modulators have a raised sine intensity modulation characteristic. A convenient linear amplitude modulator is therefore not available.

An amplitude modulated heterodyne coherent detection transmission system would be expected to have similar advantages of linewidth insensitivity and improved receiver sensitivity to an intensity modulated coherent detection system.

2.4.3 Frequency modulation

Frequency modulation is extensively used in radio communication systems because it enables modulated signal bandwidth to be traded against improved received signal-to-noise ratio. Similar advantages are available in the optical domain and are attractive since the available transmission bandwidth is large. The main practical difficulty has been to obtain an optical source capable of uniform FM response up to microwave frequencies, as discussed in Section 2.3. However, this problem has now largely been overcome by the development of the reverse-bias quantum well laser tuning technique [17].

The fundamental limitation to signal-to-noise ratio in an optical FM (OFM) system arises from the finite laser linewidth, given by:

$$SNR = \frac{\pi \Delta f^2}{2 f_m \delta f_l} \tag{2.18}$$

where Δf is the peak frequency deviation, f_m is the modulating frequency and δf_l is the source laser FWHM linewidth. For the reverse-biased quantum well tuned laser of [17] Δf was 2 GHz and δf_l 50 kHz, giving a signal-to-noise ratio of 61 dB (141 dB Hz) for a modulating frequency of 100 MHz. Commercially available semiconductor lasers would give considerably inferior performance.

Recovery of the modulating signal requires heterodyne detection followed by a suitable FM discriminator. An elegant approach to this problem is to use an heterodyne OPLL with output taken from the voltage controlled oscillator (VCO) laser control terminal. Figure 2.3 shows this arrangement.

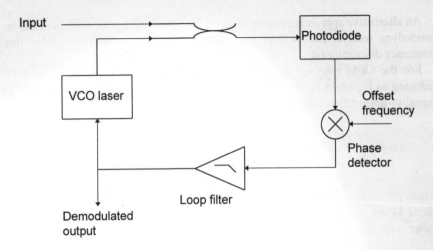

Figure 2.3 *OFM demodulator using optical phase lock loop*

Whilst the OPLL is conceptually straight-forward, realisation presents formidable technical challenges. First, the difference frequency between master and slave lasers must be set to obtain a difference frequency within the bandwidth of the photodetector. This is not difficult with diode-pumped crystal lasers. However, the temperature and current dependence of frequency for semiconductor lasers requires active temperature and current control techniques to obtain a medium term (hour) stability of the difference frequency of about 250 MHz [20].

The second challenge arises from the linewidth of the laser sources used. The OPLL must correlate the phase noise components contributing to the linewidths of the two sources in order to generate a high spectral purity difference signal, matching the supplied offset frequency. Large laser linewidths require wide loop bandwidths to achieve this. The loop bandwidth is limited by three main factors:

(i)   Signal propagation delay around the loop

(ii)   The FM frequency response of the slave laser

(iii)   The bandwidth of the loop filter and transformed bandwidth of other loop components.

For the OFM demodulation application the loop bandwidth must be greater than the base bandwidth of the signal to be demodulated. Design rules for OPLLs have been formulated [4] and in practice signal propagation delay limits base bandwidth to a few hundred MHz for loops constructed with discrete components [20,21].

An alternative approach, capable of wide base bandwidth operation, is to use an heterodyne receiver followed by a limiting amplifier and microwave delay line frequency discriminator.

For the OFM receiver the carrier-to-noise ratio is given by Equation 2.16 reducing to Equation 2.17 for shot-noise-limited reception. The improvement in signal-to-noise ratio resulting from the use of FM is:

$$SNR = 3\left(\frac{\Delta f}{f_m}\right)^3 CNR \tag{2.19}$$

where $\Delta f \gg f_m$. For $\Delta f = 2$ GHz and $f_m = 100$ MHz the improvement would be about 44 dB. This is a powerful argument for the use of OFM in wide dynamic range systems.

2.4.4 Phase modulation

Phase modulation places stringent requirements on optical source linewidth. The signal-to-noise ratio limit due to source linewidth, assuming white Gaussian frequency noise broadening, is given by :

$$SNR = \frac{\pi \Delta \phi_m^2}{\delta f_l \int\limits_{B_{min}}^{B_{max}} f^{-2} df} \tag{2.20}$$

where $\Delta \phi_m$ is the RMS phase deviation and the signal bandwidth limits are given by B_{max} and B_{min}. For an RMS phase deviation of 45 degrees and bandwidth from 100 kHz to 100 MHz, a laser of 50 kHz linewidth would give a signal-to-noise ratio of only 12 dB. For this reason phase modulation analogue transmission systems have received relatively little attention. A method of overcoming the phase noise sensitivity by using a common laser in an heterodyne arrangement has been studied by Kalman and Kazovsky [22]. However, since this requires the local oscillator signal to be transmitted with the signal it would be difficult to apply to long links.

2.5 Experimental optical frequency modulation system

As an example of a coherent analogue transmission system a brief description of an OFM link with base bandwidth 1 GHz will be given [23]. Figure 2.4 is a block diagram of the system.

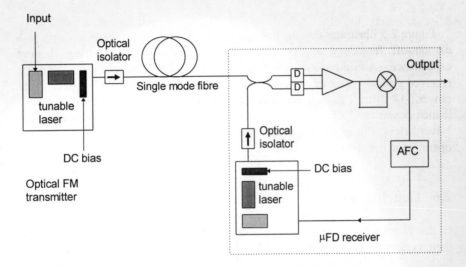

Figure 2.4 *Experimental optical frequency modulation transmission system.*

The source laser was a reverse-biased quantum-well-tuned GaAs/AlGaAs laser, having an external cavity length of 15 mm [17]. The local oscillator laser was a grating-tuned external-cavity laser of similar cavity length, which could be tuned over a wavelength range of 5 nm for channel selection in a multi-channel system. Linewidths of less than 100 kHz were achieved for both lasers.

A peak deviation of 2 GHz was chosen to give a good signal-to-noise ratio, while avoiding the danger of mode hopping in the source laser. An IF of 10 GHz was used to allow adequate demodulation linearity. The receiver used a 20 GHz bandwidth Schottky barrier photodiode, followed by a delay line discriminator for baseband signal recovery.

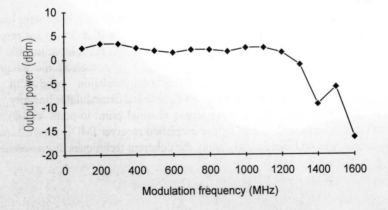

Figure 2.5 *OFM transmission system frequency response*

Figure 2.5 illustrates that the link transmission response was uniform with a -3 dB cut-off frequency of 1.3 GHz due to the source laser tuning element capacitance. A mid-band signal-to-noise ratio of 120 dB.Hz was achieved for a received optical power of -27 dBm, limited by low local oscillator laser launch power (-18 dBm). Calculations indicate that increasing the local oscillator laser launch power to -3 dBm would improve the signal-to-noise ratio to 140 dB.Hz. Nevertheless, the result achieved was some 20 dB better than an IMDD link operating at the same received power [24].

2.6 Conclusions

In this chapter coherent analogue transmission systems have been analysed and their performance compared with direct detection systems. The availability of high quality optical amplifiers for the 1550 nm optical fibre transmission window makes it possible to realise high quality IMDD transmission systems at that wavelength. However, the capability of the coherent systems to select between many channels present on the same optical fibre makes them attractive in applications such as multiple antenna remoting. They also enable wide dynamic range transmission systems to be realised at wavelengths where high quality optical amplifiers are not available.

Of the modulation techniques available for coherent systems, intensity modulation enables simple, linewidth-insensitive links to be constructed. Frequency modulation offers the potential for modulated bandwidth and signal-to-noise ratio trade-off at the expense of somewhat more exacting linewidth requirements. A key requirement in such systems is a source laser having uniform FM frequency response. This requirement can be met using the reverse-biased quantum well laser tuning technique.

For the future there is considerable scope for component development, leading to improved transmission system performance. Particular priorities for research include source modulator linearisation, narrow linewidth monolithic tunable laser development, the monolithic integration of OPLLs in optoelectronic integrated circuit (OEIC) format to achieve wide demodulation bandwidth and improvements to receiver electronics to give enhanced demodulator linearity.

It is anticipated that whilst many single channel point-to-point transmission system requirements will be met by pre-amplified receiver IMDD systems, multi-channel networks will increasingly apply the coherent techniques discussed here.

2.7 References

1. Abbas, G.L., Chan, V.W.S. and Lee, T.K.: 'Local-oscillator excess noise suppression for heterodyne and homodyne detection', *Opt. Lett.*, 1983, **8**, pp.419-421

2. Chraplavy, A.R.: 'Non-linear effects in optical fibers', in *Topics in Lightwave Transmission Systems*, ed. Li, T., (Academic Press, San Diego, 1991), pp.267-295

3. Okai, M. and Tsuchiya, T.: 'Tunable DFB lasers with ultra-narrow spectral linewidth', *Electron. Lett.*, 1993, **29**, pp.349-351

4. Ramos, R.T. and Seeds, A.J.: 'Delay, linewidth and bandwidth limitations in optical phase-locked loop design', *Electron. Lett.*, 1990, **26**, pp.389-391

5. Walker, N.G. and Walker, G.R.: 'Polarisation control for coherent communications', *J. Lightwave Technol.*, 1990, **LT-8**, pp.438-458

6. Glance, B.: 'Polarisation independent optical receiver', *J. Lightwave Technol.*, 1987, **LT-5**, pp.274-276

7. Morkel, P.R., Cowle, G.J. and Payne, D.N.: 'Travelling-wave erbium fibre ring laser with 60 kHz linewidth', *Electron. Lett.*, 1990, **26**, pp.632-634

8. Williams, K.J., Dandridge, A., Kersey, A.D., Weller, J.F., Yurek, A.M. and Tveten, A.B.: 'Interferometric measurement of low-frequency phase noise characteristics of diode laser-pumped Nd:YAG ring laser', *Electron. Lett.*, 1989, **25**, pp.774-776

9. Williams, K.J., Goldberg, L., Esman, R.D., Dagenais, M. and Weller, J.F.: '6-34 GHz offset locking of Nd:YAG 1319 nm non-planar ring laser', *Electron. Lett.*, 1989, **25**, pp.1242-1243

10. Kitagawa, T., Hattori, K., Shimizu, M., Ohmori, Y. and Kobayashi, M.: 'Guided-wave laser based on erbium-doped silica planar lightwave circuit', *Electron. Lett.*, 1991, **27**, pp.334-335

11. Brinkmann, R., Sohler, W. and Suche, M.: 'Continuous-wave erbium-diffused LiNbO$_3$ waveguide laser', *Electron. Lett.*, 1991, **27**, pp.415-416

12. Koch, T.L. and Koren, U.: 'Semiconductor lasers for coherent optical fibre communications', *J. Lightwave Technol.*, 1990, **LT-8**, pp.274-293

13. Illek, S., Thulke, W. and Amann, M.C.: 'Co-directionally coupled twin-guided laser diode for broadband electronic wavelength tuning', *Electron. Lett.*, 1991, **27**, pp.2207-2209

14. Kim, I., Alferness, R.C., Buhl, L.L., Koren, U., Miller, B.I., Newkirk, M.A., Young, M.G. and Koch, T.L.: 'Broadly tunable InGaAsP/InP vertical coupler filtered laser with low tuning current', *Electron. Lett.*, 1993, **29**, pp.664-666

15. Idler, W., Schilling, M., Baums, D., Laube, G., Wunstel, K. and Hildebrand, O.: 'Y Laser with 38 nm tuning range', *Electron. Lett.*, 1991, **27**, pp.2268-2270

16. Ogita, S., Kotaki, Y., Matsuda, M., Kuwahara, Y., Onaka, H., Miyata, H. and Oshikawa, H.: 'FM response of narrow linewidth, multi-electrode, λ/4 shift DFB laser', *IEEE Photonics Technol. Lett.*, 1990, **2**, pp.165-166

17. Cai, B., Seeds, A.J. and Roberts, J.S.: 'MQW tuned semiconductor lasers with uniform frequency response', *IEEE Photonics Technol. Lett.*, 1994, **6**, pp.496-498

18. Fong, T., Sabido IX, D.J.M. and Kazovsky, L.G.: 'Linewidth insensitive coherent AM analog optical links using semiconductor lasers', *IEEE Photonics Technol. Lett.*, 1993, **4**, pp.469-471

19. Seeds, A.J. and Cai, B.: 'Optical FM for wide dynamic range links', *Workshop on Microwave Opto-electronics, European Microwave Conf.*, 1991, Stuttgart, pp.64-70

20. Ramos, R.T. and Seeds, A.J.: 'Fast heterodyne optical phase-lock loop using double quantum well laser diodes', *Electron. Lett.*, 1992, **28**, pp.82-83

21. Gliese, U., Neilsen, T.N., Lintz Christensen, E., Stubkjaer, K.E. and Broberg, B.: 'A wideband heterodyne optical phase-locked loop for generation of 3-18 GHz microwave carriers', *IEEE Photonics. Technol. Lett.*, 1992, **4**, pp.936-938

22. Kalman, R.F. and Kazovsky, L.G.: 'Demonstration of an analog heterodyne interferometric phase-modulated link', *IEEE Photonics. Technol. Lett.*, 1994, **6**, pp.1271-1273

23. Cai, B. and Seeds, A.J.: 'Optical frequency modulation link for microwave signal transmission', *IEEE MTT-S Digest*, 1994, San Diego, pp.163-166

24. Seeds, A.J.: 'Microwave opto-electronics' *Optical and Quantum Electron.*, 1993, **25**, pp.219-229

Chapter 3

Analogue intensity modulation for optical fibre video transmission

J. M. Senior and D. T. Lambert

3.1 Introduction

The widespread deployment of analogue optical fibre links started in the late 1980s with their incorporation as fibre trunk lines into coaxial cable-based CATV networks, particularly in North America. These coaxial cable television networks provide the multichannel amplitude modulated-vestigial sideband (AM-VSB) modulation over the frequency range from 50 to 88 MHz and from 120 to 550 MHz, whilst the band from 88 to 120 MHz is reserved for frequency modulated (FM) radio broadcast. Vestigial sideband amplitude modulation was employed within the CATV networks as it exhibited the good low frequency baseband characteristics of double-sideband amplitude modulation whilst conserving bandwidth, and hence it has been widely utilised for the electrical transmission of TV and similar analogue signals.

The translation of baseband signals on to an electrical subcarrier prior to frequency multiplexing many channels (over 80 in the case of some AM-VSB for coax-based CATV networks) into a composite signal prior to intensity modulating an optical source may be referred to as subcarrier intensity modulation. Moreover the baseband video signals can be modulated on to radio frequency carriers using FM and PM (phase modulation) as well as AM. In addition, there has been substantial interest in the use of microwave frequency rather than radio frequency subcarriers, which when employed is usually called subcarrier multiplexing (SCM) [1,2]. This strategy may prove very powerful for the delivery of multiple video channels within future optical fibre distribution networks [3,4].

Although the multiplexing of many channels is of prime interest for CATV network provision, this chapter primarily deals with the analogue techniques which can be employed for intensity modulation of a single video channel to enable optical fibre transmission. It is clear, however, that the multiplexing capability of the various modulation strategies is an important factor. Nevertheless the simplest form of analogue modulation for optical fibre communications, which leaves the signal in the baseband, is direct intensity modulation of the optical source. Furthermore, pulse analogue techniques, where a sequences of pulses are used as the carrier for the video signal, may also be utilised. In this case an appropriate parameter associated with the pulse train, such as the pulse width, pulse position or pulse frequency, is electrically modulated by the baseband signal prior to intensity modulation of the optical source.

Another major factor (aside from the multiplexing capability) in the comparison of the various analogue intensity modulation strategies concerns the signal, or more precisely the carrier-to-noise ratio obtained, at the optical receiver. Hence this element of the optical fibre video transmission system is discussed and analysed in Section 3.2. This is followed in Section 3.3 with consideration of the basic analogue modulation methods; namely, direct intensity modulation, double sideband intensity modulation (rather than vestigial sideband), subcarrier phase modulation and subcarrier frequency modulation. Pulse analogue techniques are then dealt with in Section 3.4 prior to a summary and comparison of the modulation methods in Section 3.5. Two practical system implementations are described in Section 3.6. The implementations outlined are for subcarrier FM and a modified pulse frequency modulation technique known as square wave frequency modulation. Finally, in Section 3.7 a brief discussion on the evaluation of modulation systems for video transmission is provided.

3.2 Optical receivers

An optical receiver consists of an optical detector and a post-detection circuit to demodulate the original signal. Modulation of the transmission device has not been considered as the transmission system is not generally affected by the applied intensity modulation format, provided that the bandwidths employed are not close to the transmission device limits. This is not the case for optical detectors where the demodulation process is fundamentally affected by the dynamic range and frequency response of the optical detector. Three types of optical detectors are generally available for transmission systems; photoconductors, positive-intrinsic-negative (PIN) photodiodes and avalanche photodiodes (APDs).

Photoconductors [5] trade off dynamic range against frequency response. Although they offer high efficiency (90%) this does not provide any significant advantages over the other two alternatives. APD and PIN devices each have applications where they are the optimum solution and neither can be completely discarded in favour of the other. There is also a choice of what to use for the photodiode pre-amplifier which immediately follows the device and again each has particular optimised applications. The pre-amplifier can be either a high input impedance, low input impedance or transimpedance configuration [6]. The low impedance amplifier offers the simplest implementation at some noise penalty, whereas the high impedance offers the greatest sensitivity but can be complex to implement. The transimpedance device has moderate implementation difficulty and performance approaching that of the high impedance configuration. Each of the pre-amplifier structures can be combined with the PIN and APD devices thus giving six possible configurations. The high impedance front end and PIN combination are now analysed and the results applied to the comparison of modulation methods. Expressions for other combinations may be derived in the same manner [7] but will not be considered here.

Figure 3.1 shows the general configuration of a PIN photodiode and a high input impedance FET pre-amplifier. The FET amplifier offers a major improvement over the low input impedance device as the photodiode is a current source detector. For a given photocurrent, then, increasing the pre-amplifier input impedance will increase the signal voltage from the photodiode. This ensures that a larger signal voltage is present at the input of the pre-amplifier before the addition of the pre-amplifier noise [8]. One aspect of the PINFET combination which can cause implementational difficulties is capacitance in the pre-amplifier input circuit. This arises from several sources and is represented as C_t in Figure 3.1. In combination with the resistor R_b it forms an integrating circuit with corner frequency $1/2\pi R_b C_t$ and hence to compensate for this reduced frequency response an equalising differentiator must be included in the receiver circuitry.

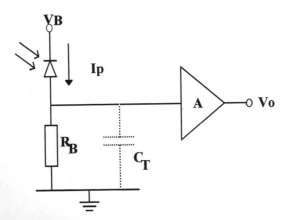

Figure 3.1 *High impedance pre-amplifier configuration*

The major noise sources arise from the photodiode, its bias resistor and the pre-amplifier gain stages. It is only possible to provide an accurate estimate of noise for a specific system configuration; however, the principle may be demonstrated by considering a particular implementation such as that shown in Figure 3.2. This is a typical circuit which could be employed in a practical implementation and comprises a PIN photodiode with FET pre-amplifier followed by an FET cascode gain stage, equalisation circuit and post-amplifier.

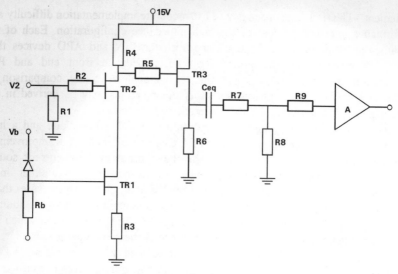

Figure 3.2 *Photodiode, pre-amplifier and secondary amplifier in a practical configuration*

Four noise sources are considered, the first of which is photodiode noise, N_P, made up of thermal noise from the photodiode equivalent series resistance and shot noise from the signal current and dark current which may be written as [9,10]:

$$N_P = \frac{2\,KT R_s}{|Z_i|^2} + e\left(I_p + I_d\right)$$

(3.1)

where K is Boltzmann's constant, T is temperature, R_S is the photodiode equivalent series resistance, e is the electronic charge, I_P is the photodiode signal current, I_d is the photodiode dark current and Z_i is the photodiode input impedance. This input impedance is given by:

$$|Z_i|^2 = \frac{R_B^2}{1 + \left(2\pi C_T R_B f\right)^2}$$

(3.2)

where R_B is the photodiode bias resistance, C_T is the total photodiode capacitance (including packaging, junction and stray contributions) and f is frequency.

The second noise source is the FET pre-amplifier in the configuration shown in Figure 3.2. It gives rise to both thermal channel noise and shot noise produced by the gate leakage current together with $1/f$ noise [11]. The total transistor noise N_T can therefore be stated as:

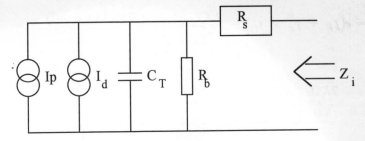

Figure 3.3 *Equivalent small circuit diagram of a photodiode*

$$N_T = eI_G + \frac{2KT\,\Gamma}{|Z_i|^2 g_M}\left(1 + \frac{f_B}{f}\right)$$

(3.3)

where I_G is the FET gate leakage current, g_M is the FET transconductance, f_B is the $1/f$ break frequency and Γ is the gamma function which is a measure of the correlation between the noise sources which make up channel noise.

Furthermore a third noise source is the thermal noise N_R generated by the photodiode bias resistor R_B and the FET bias resistors R_3 and R_4, which is given by [12] :

$$N_R = \frac{2\,KT}{R_B} + \frac{2\,KT}{|Z_i|^2 g_M^2}\left(\frac{1}{R_3 + R_4}\right)$$

(3.4)

Finally the noise contribution from the post-amplifier N_{amp} gives rise to the following noise term [13]:

$$N_{amp} = \frac{25\,KT\left(F_N - 1\right)}{R_A}$$

(3.5)

where F_N is the post-amplifier noise figure and R_A is the input impedance of this additional amplifier which is equal to 50 Ω.

Although Equations 3.1 to 3.5 accurately describe all the major noise sources associated with the receiver, the effect of these sources on the received signal depends on their location in the receiver circuit. The equations must therefore be modified in order to model the noise at the photodiode so that the signal at the photodiode can be used to estimate the overall signal-to-noise ratio (SNR). Correction factors can be calculated by using superposition theory [14] and hence applying the appropriate correction factors to Equations 3.1 to 3.5 allows them to be combined to give an overall noise function N_0 following:

$$N_0 = e(I_P + I_d + I_G) + \frac{2\,KT}{R_B} + \frac{25\,KT\,(F_N - 1)}{\left(g_M(R_3 + R_4)\,R_A\left(\dfrac{C_C}{C_T}\right)\right)^2}$$

$$+ \frac{\dfrac{2KTG}{g_M}\left(1 + \dfrac{f_B}{f}\right) + \dfrac{2KT}{g_M^2}\left(\dfrac{1}{R_3 + R_4}\right) + 2KTR_s\left(\dfrac{C_J}{C_T}\right)^2}{\dfrac{R_B^2}{1 + (2\pi C_T R_B)^2 f^2}} \qquad (3.6)$$

where C_J is the photodiode junction capacitance and C_C is the capacitance on the input of the following amplifier.

Equation 3.6 can now be used to estimate the total noise current [5], $\overline{i_N^2}$, hence:

$$\overline{i_N^2} = \int_{-f_H}^{+f_H} N_0 \, df - \int_{-f_L}^{+f_L} N_0 \, df \qquad (3.7)$$

where f_H and f_L are defined by the receiver and filter bandwidths which may be either low pass filter or band pass filter giving rise to slightly different total noise functions. It should be noted that f_L cannot be set to zero because of the $1/f$ noise term in Equation 3.6. For convenience the result of the above evaluation can be written :

$$\overline{i_N^2} = 2\,B_S\left(e(I_P + I_d) + \frac{2\,KT}{R_B}\right) + N_A \qquad (3.8)$$

where B_S is the noise equivalent bandwidth and N_A represents the total noise current due to the signal amplification which is given by:

$$N_A = 2\,B_S\left(e\,I_G + \frac{2\,KT}{R_B^2}\left(\frac{G}{g_M} + R_S\left(\frac{C_J}{C_T}\right)^2 + \frac{1}{g_M^2(R_3 + R_4)}\right)\right)$$

$$+ \frac{50\,KT\,B_S(F_N - 1)}{g_M\,R_A(R_3 + R_4)\left(\dfrac{C_C}{C_T}\right)^2} + \frac{2\,KT\,G\,f_B}{g_M\,R_B^2}\,In\left(\frac{f_H}{f_L}\right) + (2\pi C_T)^2\,\frac{4\,KT\,G\,f_B}{g_M}\left(\frac{B_S^2}{2}\right)$$

$$+ 4\,KT\,(2\pi C_T)^2\left(\frac{G}{g_M} + R_S\left(\frac{C_J}{C_T}\right)^2 + \frac{1}{g_M^2(R_3 + R_4)}\right)\left(\frac{B_S^3}{3}\right) \qquad (3.9)$$

Note that this assumes an ideal low pass filter, for a band pass filter B_S^2 is replaced by $f_H^2 - f_L^2$ and B_S^3 is replaced by $f_H^3 - f_L^3$.

Furthermore the total signal current, $\overline{i^2}_s$, is given by [1] :

$$\overline{i_S^2} = \frac{(M_O I_P)^2}{2}$$

(3.10)

where M_O is the optical modulation depth. Combining Equations 3.8 and 3.10 allows an expression for the carrier-to-noise ratio (CNR) for the incoming signal to be obtained [15] :

$$CNR = \frac{(M_O I_P)^2}{4 B_S \left[e(I_P + I_d) + \dfrac{2KT}{R_B} \right] + 2 N_A}$$

(3.11)

This expression is used to compare different analogue modulation intensity systems in terms of their expected signal-to-noise ratio performance. From this analysis, the carrier to noise ratio for a PIN FET optical receiver can be stated [11]:

$$CNR = \frac{(M_o I_p)^2}{2 (2eBI_p + N)}$$

(3.12)

where M_O = optical modulation depth, I_p = photodiode current, B is F_H-F_L for subcarrier system or F_H for baseband system and N is a noise function which can be obtained from Equation 3.8.

3.3 General analogue modulation methods

Equation 3.12 is a statement of the optical received power and the receiver and amplifier noise power, assuming a symmetrical double-sided noise distribution around a carrier frequency. N in this case must be calculated using f_c - B and f_c + B as F_L and F_H with the noise bandwidth, B_n , becoming $2B$. For the modulation methods double sideband intensity modulation (DSB-IM), sub-carrier frequency modulation (SCFM) and phase-modulated intensity modulation (PM-IM), the SNR formula will always contain this factor, with a suitable choice of frequency limits.

3.3.1 Direct intensity modulation (DIM)

For DIM the SNR is equal to the CNR [16], however, there is a covention for video SNR [17] which is shown in the following formula;

$$\text{SNR}_{dB} = 10 \log_{10} (3.92 \text{ CNR}) + w_f \qquad (3.13)$$

where w_f is the video weighting function and the factor of 3.92 arises from the convention of using peak luminace signal and rms baseband video noise.

Although DIM is obviously not a subcarrier method, Equation 3.13 is correct if suitable frequency limits are chosen for F_H and F_L, and it is easily comparable to similar equations obtained for other methods. Moreover it is possible to obtain a plot of the receiver sensitivity against the SNR (Figure 3.4) by substituting values of I_p into Equation 3.13 and solving for the SNR. Typical values for the parameters introduced in the derivation above are included in Appendix 3.1.

The CCITT limit for transmission of conventional television is 53 dB [17] and, from Figure 3.4, it can be seen that this corresponds to a receiver sensitivity of -37.5 dBm for DIM transmission. The source linearity requirement is such that, for LED implementations, the remaining dynamic range is insufficient for realistic applications. Injection laser sources require precise biasing and the optical modulation depth is limited by the laser threshold current [18]. In its favour DIM has a low transmission bandwidth (6 MHz) and has been readily demonstrated with the additional advantage that the implementation of a practical circuit is of low complexity. For example multiplexed DIM systems (DSB-IM) have been extensively used for conventional cable TV transmission in the United States [19].

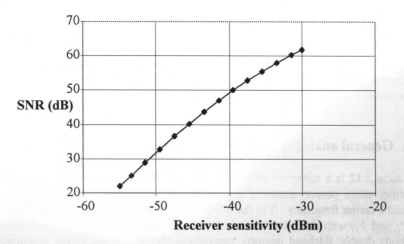

Figure 3.4 *SNR as a function of receiver sensitivity for DIM*

3.3.2 Double sideband intensity modulation (DSB-IM)

Just as DIM is perhaps the simplest form of direct modulation, DSB-IM is one of the easiest subcarrier techniques to implement. The frequency content of the original signal is shifted in the frequency domain but the modulation information is still carried by the amplitude of the signal at a point in time. This distinguishes it from subcarrier frequency modulation which utilises the instantaneous frequency of the transmitted signal to convey the modulation data. Because amplitude data is the information medium, this method also requires a high degree of source and detector linearity. Indeed as the sideband information is at a reduced amplitude then the linearity requirement is higher than that of DIM.

Taking into account the increased bandwidth requirement (approximately 10 MHz), the transmission performance of this modulation method can be characterised by using Equation 3.12. Employing this expression the SNR can be stated [20] :

$$\text{SNR}_{dB} = 10 \log_{10} (3.92 \ (M_{DSB})^2 \ \text{CNR}) + w_f \qquad (3.14)$$

where M_{DSB} is the electrical DSB modulation depth.

Simplistically this appears to indicate a difference in SNR to DIM of :

$$\text{SNR}_{(DIM)-(DSB)} = 10 \log_{10} (M_{DSB})^2 \qquad (3.15)$$

Equation 3.15 suggests that there is a 3 dB penalty associated with DSB-IM modulation in comparison to DIM, assuming a practical electrical modulation depth (M_{DSB}) of 70%. If the CNR expression is examined, however, a further difference is noted in the noise term N. Hence for a more precise comparison it is necessary to re-calculate the receiver sensitivity incorporating the difference in bandwidth. With an M_{DSB} of 0.7 this yields the data shown in Figure 3.5 indicating a receiver sensitivity of -33.0 dBm at 53 dB SNR, yielding a degradation of approximately 4.5 dB compared to DIM.

A suppressed carrier version of this technique exists incorporating single sideband operation to reduce the required transmission bandwidth and improve the SNR performance. Single sideband operation is, however, not suitable for video transmission due to its poor low frequency response which causes differential phase and gain penalties [15].

3.3.3 Subcarrier phase modulation (PM-IM)

In terms of receiver cost, subcarrier phase modulation is one of the least economic methods of analogue modulation. This is due to the increased complexity of the transmitter and receiver necessary to decode the instantaneous phase of the signal. The modulated signal bandwidth is determined from Carson's rule:

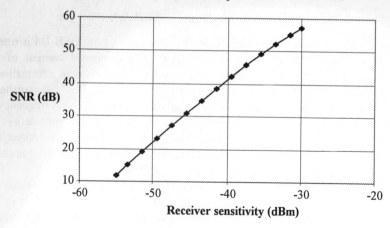

Figure 3.5 *SNR as a function of receiver sensitivity for DSB-IM*

$$B_M = 2 \ (M_{PM}+1) \ B \qquad\qquad (3.16)$$

where M_{PM} is the electrical phase modulation index, B is the baseband signal and B_M the modulated signal bandwidth.

For practical values of M_{PM}, PM-IM requires a high transmitted signal bandwidth compared to most other systems. Also, as the technique relies on setting the instantaneous phase of the subcarrier proportional to the modulated signal, there will be jitter-related penalties associated with the operation of this system at the required data rates. These difficulties may be compounded when multiplexing is employed. If lower values of M_{PM} are used to reduce the transmission bandwidth, then performance penalties will be incurred unless phase resolution is improved. This is particularly undesirable in video transmission applications.

Using the CNR expression quoted earlier in Equation 3.12 an SNR equation can be formulated [21]:

$$SNR = 10 \ \log_{10} (3.92 \ (M_{PM})^2 \ CNR) + w_f \qquad\qquad (3.17)$$

Equation 3.17 is graphically represented in Figure 3.6, indicating a receiver sensitivity of -34.0 dBm at 53 dB SNR. It should be noted that no account has been taken of threshold effects as they occur at an SNR well below that utilised in normal operation (approximately 30 dB SNR for $M_{PM} = 3$) [22].

3.3.4 Frequency modulation (FM)

FM coding is traditionally associated with improvements in the radio transmission of audio signals. This is because the audio range covers a limited frequency band, and the additional bandwidth required for frequency-modulated transmission could be accommodated within the allowed spectrum space. For aerial television transmission this is not possible because of the restricted frequency band allocated to each TV channel. However, in optical fibre applications, band allocations have not been applied and the transmission medium has a large amount of frequency headroom which is limited largely by cost.

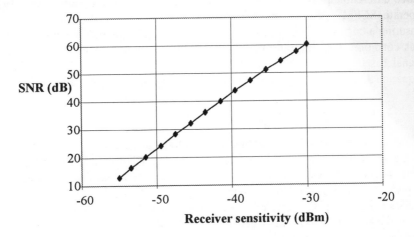

Figure 3.6 *SNR as a function of receiver sensitivity for PM-IM*

Theoretically, baseband FM transmission is feasible but this implies DC operation which is not practical as the frequency distribution is double sided and asymmetric overlapping of the low frequencies would occur. FM subcarrier techniques are therefore normally utilised. Unlike other subcarrier techniques, the carrier frequency is only present when no modulating input is applied. This is because the demodulation process utilises instantaneous frequency differences to recover the original signal information and does not require a fixed frequency reference. Subcarrier FM (SCFM) is a very well developed modulation method which has been extensively used for conventional TV transmission, particularly in cabled systems [23]. Optical source intensity linearity is not as important as for DIM, DSB-IM and PM-IM but a DC offset on the waveform may affect the demodulation of the signal leading to distortion. The modulator and demodulator linearity, rather than the optical source linearity, will control the overall system performance [24].

As in the case of PM-IM there is a bandwidth penalty associated with FM transmission quantified by Carson's rule:

$$B_M = 2 \ (M_{FM} + 1) \ B \qquad (3.18)$$

where M_{FM} is the electrical FM modulation depth and B_M and B are as described in Equation 3.16. Narrowband FM techniques utilise a frequency deviation index

(M_{FM}) of less than 1, but there are numerous problems and subsidiary requirements associated with this method [25]. For example, the deviation index is a measure of the frequency resolution (and therefore the demodulated signal amplitude resolution) of the transmitted signal. As such it effectively controls the linearity resolution which is reflected in the direct relation between the deviation index, the video differential phase and the video differential gain. It is therefore necessary to operate SCFM systems with a frequency deviation index of one (or greater) ensuring a correspondence between the frequency and amplitude resolution. With this condition then the transmission bandwidth becomes 24 MHz compared to a signal baseband bandwidth of 6 MHz. Moreover using the CNR described in Equation 3.12, the SNR of an SCFM transmission system can be stated [17]:

$$\text{SNR} = 10 \log_{10}(11.76 \ (M_{FM})^2 \ \text{CNR}) + w_{fm} \qquad (3.19)$$

where w_{fm} is the FM video weighting function equal to 12.2 dB.

Equation 3.19 is only strictly accurate for high CNR, where the video noise spectral density is related to f^2. For low CNR the video noise loses its frequency dependence and it is therefore essentially constant with w_{fm} becoming equal to w_f. Additionally, there is a threshold effect due to the demodulation method which gives rise to an increasing SNR penalty as the CNR reduces. Both of these effects can be taken into account if Equation 3.19 is modified to :

$$\text{SNR}_{dB} = 10 \log_{10} \left(\frac{11.76 \ (M_{FM})^2 \ \text{CNR}}{1 + 24 \ M_{FM} \ \text{CNR} \exp \left(\frac{-2 \ \text{CNR}}{M_{FM} + 1} \right)} \right) + w_f \qquad (3.20)$$

where the denominator is derived from the effect of rectangular filtering on the transmitted signal [26]. For $M_{FM} = 1$, the threshold effect occurs at very low SNR (see Figure 3.7) compared to the 53 dB video performance limit, at which the receiver sensitivity is -37.5 dBm.

For the optimum SNR performance M_{FM} should be adjusted to move the threshold point to the operational SNR region. This, however, gives a much increased transmission bandwidth which is undesirable and impractical. Moreover it also has a dramatic effect on SNR if the received signal is reduced (i.e. approximately 7 dB loss in SNR for each additional dB of optical loss).

3.4 Pulse analogue techniques

Work has been undertaken utilising pulse analogue techniques with well known systems such as pulse width modulation (PWM) and pulse position modulation

(PPM) receiving significant attention [27,28]. There are a number of reasons for examining pulse analogue techniques for optical fibre applications; however perhaps the most important is the suitability of the signal for injection laser transmission. The techniques remove the necessity for a stringent optical source linearity limit and replace it with a more achievable electrical modulation linearity requirement. Additonally, PWM has been used in the transmission of conventional video signals under experimental conditions [29].

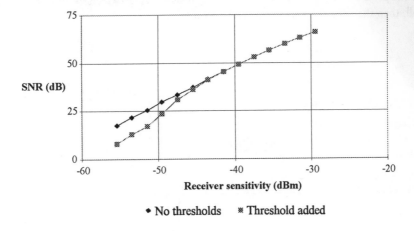

Figure 3.7 *SNR as a function of receiver sensitivity for SCFM*

Three techniques will be considered, pulse width modulation (PWM), pulse position modulation (PPM) and pulse frequency modulation (PFM). A fourth, square wave frequency modulation (SWFM), will be discussed with PFM as there is a fundamental relationship between the two systems. Other techniques exist such as pulse interval modulation (PIM) and pulse interval and width modulation (P(I+W)M). However, PIM has been shown to be inferior to PPM in terms of noise performance [30] and to date P(I+W)M has only been utilised under laboratory conditions at audio frequencies [31].

3.4.1 General noise analysis for pulse analogue techniques

PWM and PPM can be termed synchronous techniques which is to say there is a fixed time between consecutive rising or falling pulse edges. Although PFM of necessity has a variable frequency content, there is a fixed time between the rising and falling edge of a single pulse and this enables similar analysis to be used for its noise performance. Generally the three techniques can be represented in the time domain by Figure 3.8. This shows a pulse rising at time t_o, falling at time $(t_o + t_p)$ depending on the modulation method used and rising again at time $(t_o + t_s)$. Here

t_s is the time between samples and t_p is the pulse width which carries the modulation information.

For PFM, however, the above roles are reversed. At each of these times the presence of noise can affect the time that the threshold is crossed, thereby affecting the demodulated signal amplitude. The effect of this time variation can be estimated by considering Figure 3.9 which shows the modulation information as a function of time, $p(t)$, with the effect of a noise function $n(t)$ added. By assuming that the $p(t) + n(t)$ signal is parallel to $p(t)$, which is approximately true if $n(t)$ is small, then the zero crossing point $x(t_o)$ becomes:

$$x(t_o) \approx - \frac{n(t)}{p'(t_o)}$$

(3.21)

where $p'(t_o)$ is the gradient of $p(t)$ at $t = t_o$.

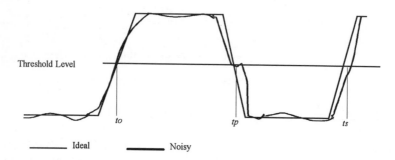

Threshold Level

to   tp   ts

———— Ideal   ———— Noisy

Figure 3.8 *General waveform for pulse analogue methods*

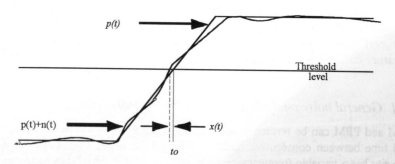

$p(t)$

Threshold level

$p(t)+n(t)$ $x(t)$

to

Figure 3.9 *The effect of noise on pulse analogue signals*

For a continuous pulse train, noise-free threshold crossings occur at :

$$t = t_o + m\, t_s$$

(3.22)

and

$$t = t_0 + t_p + m\,t_s \tag{3.23}$$

The introduction of noise, however, effectively produces noise pulses in the demodulator which can be replaced by impulse functions for low noise (narrow pulses). Hence the noise function can be represented by [32]:

$$x_n(t) = \sum_{m=-\infty}^{\infty} -[x(t_0 + m\,t_s)\delta(t - t_0 - m\,t_s) - x(t_0 + t_p + m\,t_s)\delta(t - t_0 - t_p - m\,t_s)] \tag{3.24}$$

where $\delta(t)$ is the unit impulse function.

The gradient at the threshold point is generally taken to be constant, which simplifies Equation 3.24 to:

$$x_n(t) = \left(\frac{n(t)}{p'(t_0)}\right) \sum_{m=-\infty}^{\infty} [\delta(t - t_0 - m\,t_s) - \delta(t - t_0 - t_p - m\,t_s)] \tag{3.25}$$

It can be considered physically as the noise on the carrier signal $n(t)$, combined with the noise on the modulating signal, which is made up of two periodic impulse trains delayed by the pulse width t_p. This periodicity allows Equation 3.25 to be separated into real and imaginary components:

$$x_n(t) = \left(\frac{n(t)}{t_s\,p'(t_0)}\right) \sum_{k=-\infty}^{\infty} \left(1 + \exp\left(-j2\pi k\frac{t_p}{t_s}\right)\right) \exp\left(j2\pi k\frac{t_p}{t_s}\right) \tag{3.26}$$

In order to evaluate the frequency performance of the system, the Fourier transform of Equation 3.26 must be taken, hence:

$$x_n(f) = \left(\frac{f_s}{p'(t_0)}\right) \sum_{k=-\infty}^{\infty} [1 + \exp(-j2\pi f_s k t_p)]\, N(f - kf_s) \tag{3.27}$$

where $f_s = 1/t_s$ and $N(f)$ is the Fourier transform of $n(t)$.

Now the noise spectral density $N_{sd}(f)$ can be estimated as:

$$N_{sd}(f) = |x_n(f)|^2 \tag{3.28}$$

which by substitution from Equation 3.27 becomes:

$$N_{sd}(f) = 2\left(\frac{f_s}{p'(t_o)}\right)^2 \sum_{k=-\infty}^{\infty} [1+\cos(2\pi f_s k\, t_p)]\, N_{eq}(f-kf_s)$$

(3.29)

Here $N_{eq}(f) = |N(f)|^2$ which corresponds to the noise spectral density arising from the pulse equaliser used.

In practice $N_{sd}(f)$ is largely flat as it arises from the addition of many $N_{eq}(f)$ terms at $\pm kf_s$ intervals. Also Equation 3.29 is a general case in which the receiver responds to all threshold crossings taking into account their polarity. Two special cases of the noise spectrum arise; firstly from systems which utilise all pulse edges whether rising or falling, $N_{sd}(f_p)$, and secondly from systems where either rising edges or falling edges alone are utilised. The second of these is termed the alternate crossing case represented by $N_{sd}(f_a)$. The appropriate expressions for these two situations are:

$$N_{sd}(f_p) = 2\left(\frac{f_s}{p'(t_o)}\right)^2 \sum_{k=-\infty}^{\infty} [1-\cos(2\pi f_s k\, t_p)]\, N_{eq}(f-kf_s)$$

(3.30)

and

$$N_{sd}(f_a) = 2\left(\frac{f_s}{p'(t_o)}\right)^2 \sum_{k=-\infty}^{\infty} N_{eq}(f-kf_s)$$

(3.31)

The pulse gradient $p'(t_o)$ is limited in practice by the response of the pulse equaliser used and it must be estimated in order to calculate the noise spectral density. The demodulated pulse shape $p(t)$ is therefore given by:

$$p(t) = p_r(t) * h_e(t)$$

(3.32)

where $h_e(t)$ is the equaliser impulse response, $p_r(t)$ is the pulse shape at the receiver and $*$ is the symbol for convolution. In addition, for low receiver noise the following approximation can be made [32]:

$$p'(t_o) \approx P_r'(t_o) * h_e(t)\,|_{t=t_o}$$

(3.33)

where $P_r'(t_o)$ is the derivative of the pulse shape at the receiver and $|_{t=t_o}$ represents the limit of the convolved function at $t = t_o$.

Generally the equaliser (rather than the receiver) bandwidth is the limiting factor in this expression. If the receiver response, $p_r(t)$, is rectangular, then the derivative $P_r'(t)$ will be constant across the entire frequency band of interest. Equation (3.33) then becomes:

$$p'(t_o) = M_O I_p\, \delta(t_o)\, h_e(t)\,|_{t=t_o}$$

(3.34)

where $\delta(t)$ is the unit impulse function, M_O = optical modulation depth and I_p = photodiode current. This can be approximated to [32]:

$$p'(t_o) \approx M_O I_p h_e(t_o) \qquad (3.35)$$

The final step is to normalise $h_e(t_o)$ to the equaliser bandwidth B_e, which then gives:

$$p'(t_o) = 2 \pi B_e M_O I_p h_n(t_o) \qquad (3.36)$$

where $h_n(t_o)$ will be a constant depending on the equaliser chosen. For example, if a second order Butterworth equaliser were used, $h_n(t_o)$ is optimised at 0.456 [33]. The presence of the filter will give rise to a power transfer function, which for the second order Butterworth filter is given by:

$$\left| H_e(f) \right|^2 = \frac{1}{1 + \left(\dfrac{f}{B_e} \right)^{2n}} \qquad (3.37)$$

The analysis above is applicable to all the pulsed analogue techniques considered but further development of the analysis is required for individual methods. These analyses are now considered in turn.

3.4.2 Pulse width modulation (PWM)

The PWM signal contains a baseband component in its frequency spectrum which reproduces the original signal when low pass filtering is applied (see Figure 3.10). The demodulator noise is therefore determined by the voltage output of the device used with no additional contributions from other circuit components. The noise D_n therefore follows from Equation 3.29:

$$D_n = 2 \left(\frac{V_p f_s}{p'(t_o)} \right)^2 \int_{-B}^{B} \sum_{k=-\infty}^{\infty} [1 + \cos(2\pi f_s k t_p)] N_{eq}(f - k f_s) \, df \qquad (3.38)$$

where V_p is the output signal voltage from the demodulator.

Equation 3.38 follows directly from Equation 3.29 with V_p added to take into account the amplitude of the demodulated pulse. As the noise spectral density is essentially flat this can be simplified to:

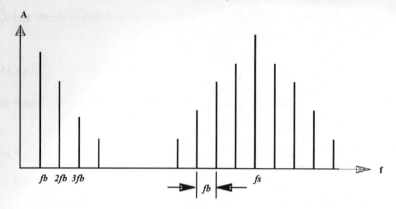

Figure 3.10 *Frequency spectrum for PWM and PPM coded signals*

$$D_n = 2 \left(\frac{V_p f_s}{p'(t_o)} \right)^2 2 B \eta_{PW}$$

(3.39)

where η_{PW} is a constant given by:

$$\eta_{PW} = \sum_{k=-\infty}^{\infty} [1 + \cos(2\pi f_s k t_p)] |H_e(kf_s)|^2 N_r(f - kf_s)$$

(3.40)

where $H_e(f)$ can be found from Equation 3.38 and $N_r(f)$ is the equaliser noise referred to the receiver input and it can be calculated using Equation 3.30. η_{PW} can be computed at any frequency within the band transmitted, although for consistency it is normally calculated at the maximum transmission frequency, B.

Having defined the demodulator noise, it only remains to calculate the signal output in order to estimate the signal-to-noise ratio (SNR). If the pulse width is sinusoidally modulated then the demodulator output is given by [25]:

$$S_D = \frac{(V_p f_s \Delta\tau)^2}{2}$$

(3.41)

where $\Delta\tau$ is the difference between the modulated and unmodulated transmission pulse width. Combining Equation 3.39 with Equation 3.41, the SNR can be stated in a similar format to the CNR referenced analogue methods:

$$SNR = 10 \log_{10}(3.92 \, (M_{PW})^2 \, PNR_{PW}) + w_f$$

(3.42)

where $M_{PW} = 2 \Delta\tau / t_s$ and is termed the PWM modulation index and PNR_{PW} is the pulse-to-noise ratio arising from Equations 3.39 and 3.41:

$$\text{PNR}_{PW} = \frac{[\pi \, M_o \, h_e(t_s) \, B_e \, t_s \, I_p]^2}{8 \, B \, \eta_{PW}}$$

(3.43)

The selection of operational values of f_s, B_e and the threshold decision level has a profound effect on the performance and transmission bandwidth of the system. For optimum performance B_e must be around the value of the highest baseband frequency, B, and the threshold decision level must be centred around 50%. The sampling frequency, f_s, however is not so readily optimised. As the sampling frequency rises the SNR increases but the transmission bandwidth increases at twice the rate. As f_s is reduced, the baseband components of the signal begin to interfere with the modulated signal causing distortion. For practical PWM systems f_s is usually at least five times the highest signal frequency and M_{PW} is around 0.7. If a 6 MHz video signal is considered, this gives a sampling frequency of around 30 MHz and a transmission bandwidth of around 45 MHz. Note that for higher values of M_{PW} this maximum frequency would rise proportionaly, becoming 60 MHz at a modulation index of 1. As for the previous analogue systems it is possible to calculate the relation between receiver sensitivity and SNR (using Equation 3.42) and this is plotted in Figure 3.11 for the above values of f_s and M_{PW}. These figures give a value of receiver sensitivity of -34.0 dBm for a 53 dB SNR. The transmission of several multiplexed PWM channels, however, is difficult because of the presence of the baseband component of the frequency spectrum. A practical system would therefore require electrical multiplexing of the signals followed by PWM coding of the multiplexed information. This would obviously entail a much larger transmission bandwidth combined with the associated implementational difficulties.

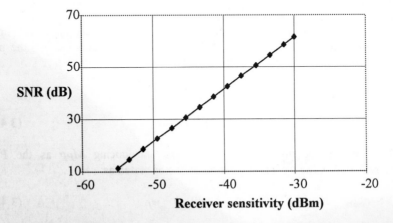

Figure 3.11 *SNR as a function of receiver sensitivity for PWM*

3.4.3 Pulse position modulation (PPM)

Pulse position modulation requires a more complex implementation than PWM or PFM, as the position of the falling edge of a pulse must be evaluated relative to the timing of a reference clock signal [34]. There is also noise associated with the recovery of the clock information from the transmitted signal. However, this factor is usually assumed to be small compared to the noise on the edges of the detected pulse train [35]. The performance of the system is therefore assumed to be limited by the temporal shift on these pulses. The reliance of the demodulation method on the reference clock does, however, simplify the noise analysis as only the magnitude of alternate threshold crossings must be considered as noise sources. The noise spectral density at the receiver can therefore be stated directly from Equation 3.31 as:

$$D_n = 2 \left(\frac{V_p f_s}{p'(t_o)} \right)^2 \int_{-B}^{B} \sum_{k=-\infty}^{\infty} N_{eq}(f - k f_s) \delta f$$

$$(3.44)$$

As in the case of PWM this can be simplified to:

$$D_n = \left(\frac{V_p f_s}{p'(t_o)} \right)^2 2 B \eta_{PP}$$

$$(3.45)$$

where η_{PP} is now given by :

$$\eta_{PP} = \sum_{k=-\infty}^{\infty} |H_e(k f_s)|^2 N_r(f - k f_s)$$

$$(3.46)$$

The output from the PPM demodulator can be represented by a similar expression to that obtained for PWM by simply replacing $\Delta\tau$ with Δt, the peak excursion in pulse position such that :

$$S_D = \frac{(V_p f_s \Delta t)^2}{2}$$

$$(3.47)$$

Combining Equations 3.45 and 3.47 and introducing M_{PP} as the PPM modulation depth, the weighted video SNR can be stated as:

$$SNR_{dB} = 10 \log_{10} (3.92 \, (M_{pp})^2 \, PNR_{pp}) + w_f$$

$$(3.48)$$

where PNR_{PP} is the PPM pulse-to-noise ratio arising from Equations 3.45 and 3.47:

$$\text{PNR}_{PP} = \frac{[\pi m_o \, h_e(t_s) \, B_e \, t_s \, I_p]^2}{4 \, B \, \eta_{PP}}$$

(3.49)

From Equation 3.48 the relation between receiver sensitivity and SNR was evaluated and is plotted in Figure 3.12, giving a receiver sensitivity of -36.0 dBm at 53 dB SNR.

The frequency content of the transmitted PPM signal is similar to that of PWM. PPM then has the same lower limit to f_s because of the presence of baseband information. This entails a sampling frequency of around 30 MHz. Typically M_{PP} is around 0.7, B_e is once more equal to B and the threshold level must be 50% for optimum performance. These values are the same as those for PWM and lead to the same transmission bandwidth of 45 MHz. The high frequencies associated with the operation of this system pose greater implementational difficulties than for PWM, especially if the transmission of multiple channels is considered.

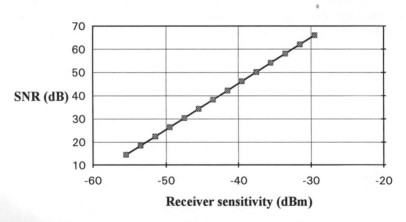

Figure 3.12 *SNR as a function of receiver sensitivity for PPM*

3.4.4 Pulse frequency modulation (PFM)

The PFM signal must be reconstructed at the receiver by passing the signal through a monostable which then enable the same receiver to be used to decode SWFM signals. The output of the monostable will then contains a baseband reconstruction of the original signal free from transmission noise and recoverable with low pass filtering. It should be noted that the PFM signal does contain a baseband reconstruction of the signal which is recoverable without utilising the monostable circuit, however, the following analysis assumes a monostable is present. As the leading and falling edges of the PFM pulse have a fixed separation, only alternate threshold crossings carry transmission noise information thus enabling Equation

3.31 to be used to characterise the noise. However, the inclusion of the monostable in the demodulator introduces a transfer function, $2\pi f \tau_{PF}$ into the expression (τ_{PF} is the width of the PFM pulse) to take the effect of the circuit into account [36]. With this factor incorporated into the noise function, the spectral density of the noise at the demodulator is given by:

$$D_n = 2 \left(\frac{2 \pi \tau_{PF} V_p f_s}{p'(t_o)} \right)^2 \int_{-B}^{B} f^2 \sum_{k=-\infty}^{\infty} N_{eq}(f - kf_s) \, \delta f$$

(3.50)

This can be simplified to:

$$D_n = \left(\frac{2 \pi \tau_{PF} V_p f_s}{p'(t_o)} \right)^2 \frac{2}{3} B^3 \eta_{PP}$$

(3.51)

where η_{PP} is given by Equation 3.46 and is applicable as both PFM and PPM only require alternate threshold crossings to be considered. The output signal from the PFM demodulator S_D is described by the following relation [37]:

$$S_D = \frac{(V_p \tau_{PF} \Delta f)^2}{2}$$

(3.52)

where Δf is the peak frequency deviation from f_s. Using the familiar FM modulation index, $\beta = \Delta f / B$, the PFM signal-to-noise ratio is given by:

$$\text{SNR}_{dB} = 10 \log_{10} (11.76 \, \beta^2 \, \text{PNR}_{PF}) + w_{fm}$$

(3.53)

where the PFM pulse-to-noise ratio PNR_{PF} is given by :

$$\text{PNR}_{PF} = \frac{[m_o \, h_e(t_s) \, B_e \, t_s \, I_p]^2}{4 B \, \eta_{PP}}$$

(3.54)

Threshold effects occur in all the pulsed analogue techniques but they occur at much lower SNRs for PWM and PPM. As the effect is likely to be significant at or around the SNR of interest (as in SCFM), analysis is necessary to provide an accurate estimate of the system performance. The threshold analysis for PFM is more complex than for SCFM but the effect can be approximated by modifying Equation 3.53 to include an extra term [38].

$$\text{SNR}_{dB} = 10 \log_{10} \left(\frac{11.76 \, \beta^2 \, \text{PNR}_{PF}}{1 + 3\pi \left(\dfrac{B_{rms}}{f_s} \right) \left(\dfrac{h_e(t_s) B_e}{B} \right)^2 R_{pn} \exp\left(\dfrac{-R_{pn}}{8} \right)} \right) + w_{fm}$$

(3.55)

where R_{pn} is given by:

$$R_{pn} = \frac{(M_o I_p)^2}{\displaystyle\int_{-\infty}^{\infty} N_{eq}(f) \, \delta f}$$

(3.56)

B_{rms} is given by [39]:

$$B_{rms} = \left(\frac{\displaystyle\int_{-\infty}^{\infty} f^2 \, N_{eq}(2\pi f) \, \delta f}{\displaystyle\int_{-\infty}^{\infty} N_{eq}(2\pi f) \, \delta f} \right)^{\frac{1}{2}}$$

(3.57)

and N_{eq} is defined as in Equation 3.29.

The SNR expression given in Equation 3.55 is equally valid for SWFM and PFM, assuming that they are demodulated using the same strategy (i.e. the received signal is used to trigger a monostable and the resultant signal is low pass filtered). Using this expression the receiver performance can be calculated and is shown in Figure 3.13, giving a receiver sensitivity of -37.5 dBm at an operating SNR of 53 dB.

3.5 Summary of modulation methods

Examining the four directly modulated techniques, DIM, DSB-IM, PM-IM and SCFM, significant differences in receiver sensitivity and transmission bandwidth are observed and these are tabulated in Table 3.1. DIM and SCFM have the highest receiver sensitivities with DIM also having the lowest transmission bandwidth. Receiver sensitivity is a particularly important parameter for optical transmission systems and the relatively poor performances of DSB-IM and PM-IM make their use unlikely. PM-IM also has the highest transmission bandwidth requirement at 48 MHz, has a complex receiver and has not been demonstrated at the necessary transmission bandwidths. Although DIM has the lowest transmission bandwidth and good optical transmission performance, it is unlikely to be used for injection laser-based applications because of the optical source linearity required. The linearity requirement for DSB-IM is also restrictive and, coupled with the 4.5 dB receiver sensitivity penalty, precludes further consideration. DIM could be used for LED based transmission systems where, because of its low bandwidth, it would probably be the optimum choice. Hence for injection laser-based systems and multi-channel applications SCFM is the preferred method.

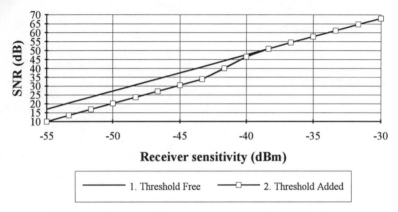

Figure 3.13 *SNR as a function of receiver sensitivity for PFM*

Table 3.1 *Summary of analogue modulation methods*

Modulation Method	Transmission Bandwidth (MHz)	Linearity	Receiver Sensitivity (dBm)	Receiver Complexity
DIM	6	High	-37.5	Low
DSB-IM	12	V.High	-33.0	Medium
PM-IM	48	V. High	-34.0	High
SCFM	24	Low	-37.5	Medium
PWM	36	Low	-34.0	Low
PPM	36	Low	-36.0	High
PFM/ SWFM	24	Low	-37.5	Low

For the pulsed analogue methods, receiver sensitivity is, once more, a major consideration and PFM and SWFM shows an advantage in this area over the other techniques. PFM and SWFM also has the lowest transmission bandwidth requirement. In terms of receiver complexity PWM, PFM and SWFM are the simplest with PFM and SWFM requiring a monostable in addition to the low pass filter used in PWM, although as previously stated PFM can be implemented without the monostable. Optical source linearity should not affect the pulse transmission unless the transmission bandwidth is limited and then only by a small amount. To explain this statement the frequency spectrum of all of the methods has harmonic content which controls the gradient of the pulse edge p' (t_0) (see Equation 3.21). If this gradient is reduced by a loss of harmonic information then any non-linearity of the source could introduce a timing offset which translates to a change in the demodulated signal amplitude. If the source non-linearity is known, it can be compensated for by adjusting the decision threshold level. Also the transmission bandwidth is centred around the sampling frequency in the frequency domain. For example, the computed values of SNR obtained for PWM assume a sampling frequency of 30 MHz and a transmission bandwidth of 45 MHz. In the generated signal, however, there are harmonics between 67.5 and 135 MHz. These would normally be filtered out and $p'(t_0)$ would be limited by the filter response. SNR improvements may be achieved where it is possible to allow these harmonics to be transmitted as they would have a significant effect on $p'(t_0)$ which has an inverse square relation to SNR.

PFM has baseband signal content up to 6 MHz and for a 24 MHz centre frequency a modulated signal between 12 to 36 MHz where once again higher harmonics are filtered out. Perhaps the most interesting case is that of SWFM where there is no baseband signal component. The centre frequency can be set lower, to 18 MHz for example, giving an unavailable band of 6 to 30 MHz, hence the bandwidth value of 24 MHz in Table 3.1. Additionally, if the actual transmitted SWFM signal is considered it can be seen that it is identical to that of SCFM. It is perhaps no surprise then that the receiver sensitivity of SWFM is identical to that of SCFM. Of the pulsed analogue methods, SWFM with its lack of baseband signal content, low receiver complexity and simpler multiplexing strategy therefore appears most appropriate for video transmission.

3.6 Examples of practical system implementations

The following section describes the implementation strategy for two analogue modulation systems, SCFM and SWFM. Some general considerations for FM system design are discussed first, followed by a description of a possible SCFM system implementation. The modifications necessary to produce an SWFM system from the SCFM implementation are then detailed and finally the methods to assess the relative performance of modulation systems are discussed.

3.6.1 FM modulator and demodulator design considerations

The most critical components of any FM-based system are the voltage-to-frequency (V-to-F) and frequency-to-voltage (F-to-V) convertors. In video applications these items take on an even higher degree of importance as any non-linearity in the modulator would be seen as an intensity change in the received picture when demodulated. Obviously both modulator and demodulator are critical as a non-linearity in either will affect the displayed signal. It is vital then that the conversions are linear over the region of interest.

The two most important operational characteristics are modulation depth and centre frequency which generally determine the choice of modulator and V-to-F devices are readily available to convert conventional bandwidth TV signals to FM format. Other considerations are the systems complexity, cost and multiplexing capability. Complexity and cost are not limiting factors associated with the modulator as one modulator could serve several local loop subscribers. The demodulator is, however, very cost-sensitive as one would be provided at each terminal. Subcarrier FM techniques are, by design, relatively simple to multiplex, requiring only a duplication of the modulator with a different centre frequency [40,41]. The two resulting signals can then be passively combined.

In order to ensure that consumer costs are kept to a minimum, the demodulator and receiver costs must be kept low as these are the most expensive system components. For FM decoding, a standard limiter-discriminator demodulator is the established signal recovery technique. The linearity of these systems is very good and devices are readily available to construct demodulators operating up to the GHz frequency range.

3.6.2 Subcarrier FM system implementation

An example of a system implementation for subcarrier FM (SCFM) is shown in Figure 3.14. The TV signal is modulated and mixed to an intermediate frequency before low pass filtering and optical transmission. At the receiver, the transmitted signal is demodulated, amplified to video levels and displayed on the monitor.

3.6.2.1 Modulation and transmission

The FM implementation can be carried out with an Avantek VTO 9050 which has a maximum available modulation depth of 20 MHz. To establish this figure an examination of the relationship between the frequency content of the modulated signal and the voltage of the input signal must be made. For the VTO 9050 a typical V-to-F response for DC voltages is shown in Figure 3.15. This characteristic demonstrates that a peak to peak signal of 2 V centred around 8 V would give a frequency output of 690 MHz $\pm$ 35 MHz. If the above signal was modulated at a low frequency (e.g. 100 kHz) then the frequency spectrum shown in Figure 3.16 would be observed. The two peaks are 70 MHz apart and the frequency content is indicative of a very high modulation index, 350 in fact.

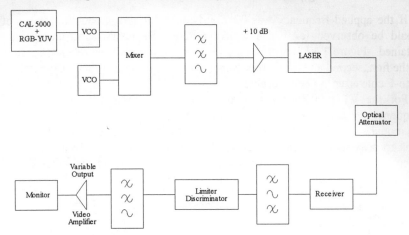

Figure 3.14 *Subcarrier FM practical implementation*

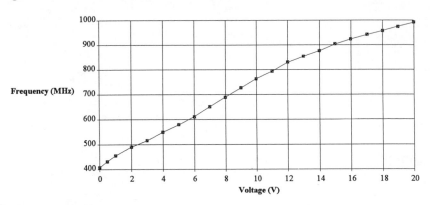

Figure 3.15 *VTO9050 Voltage-to-frequency response*

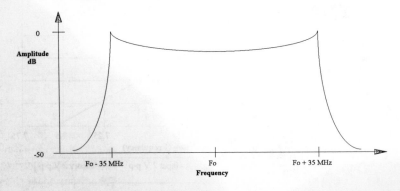

Figure 3.16 *Frequency response of a signal with β=350*

If the applied frequency was increased to 35 MHz, a modulation index of 1 would be observed for which a very different frequency spectrum would be obtained. Figure 3.17 shows this frequency spectrum with the relative amplitudes of the first, second and third order components indicating that the 3 dB point of the V-to-F converter has been reached.

For the VTO 9050 this spectrum is generally observed when the applied frequency (Fm) is only 20 MHz. The AC response is best demonstrated by plotting the modulation index against the input frequency on logarithmic scales. For the 2 V peak-to-peak signal considered above, the solid line in Figure 3.18 is the response calculated from the DC characteristic. The actual response assuming 20 MHz capability is also plotted on this figure as is the response with 1 V input.

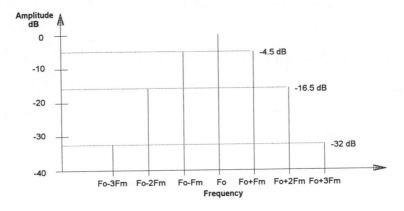

Figure 3.17 *Frequency response of a signal with $\beta=1$*

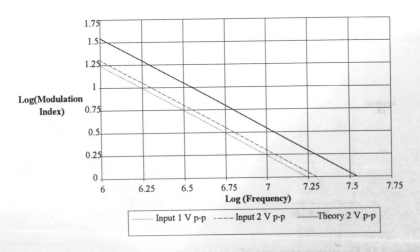

Figure 3.18 *VTO9050 Maximum useful input voltage measurement*

By comparing the theoretical 1 V and 2 V responses to the expected response, the maximum useful input voltage can be calculated, which in this case is 1.14 V, equivalent to a 20 MHz modulation depth. Any increase in the applied voltage above this value would give the same frequency response. The above discussions clearly demonstrate the capability of the VTO 9050 to encode TV bandwidth signals.

Using the VTO 9050 the circuit configuration shown in Figure 3.19 can be assembled containing two of the devices. The first is the active V-to-F converter and the second is set up with a DC input voltage to provide a fixed reference frequency with which to mix the modulated signal. If an external frequency source is used to provide the reference, then the temperature dependence of the VTO 9050 will cause it to drift relative to the more stable external source. Using a second VTO 9050 as the frequency reference will apply the correct compensation to ensure a fixed centre frequency for the mixed signal. Similarly, a potentiometer can initially be used to set the voltage for the second device but the temperature dependence of the potentiometer will be sufficiently different to that of the fixed resistor used to bias the first device that frequency drift will occur. Once the optimum settings have been determined with the potentiometer it can be removed and fixed resistor values shown in Figure 3.19 can be utilised.

For the frequency reference device only the ratio of the two resistors is important, but for the active V-to-F converter the input resistors have to be chosen to match the 75 Ω video output impedance and to give the correct ratio for the required output centre frequency. The large (2200 μF) capacitor on the active VCO input is also required in order to maintain the low frequency content (as low as 10 Hz) of the video signal. The mixer used in this example circuit is a Mini-Circuits SRA-11 device capable of operating with up to 2 GHz at its inputs and a 600 MHz intermediate frequency (IF) output. It is termed a level 7 device which requires at least +7 dBm (into 50 Ω) signal level for the local oscillator (LO) input.

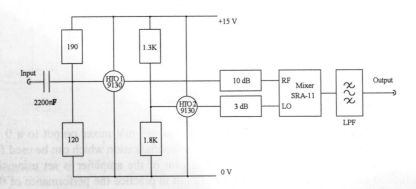

Figure 3.19 *VTO and mixer configuration*

There is also a maximum allowed signal level at the RF or signal input of +1 dBm and hence the attenuator arrangement in Figure 3.19. Finally, the centre frequency of the active VCO is 750 MHz and that of the reference device 718 MHz giving an IF of 32 MHz. The final component in the circuit is a passive low pass filter which should be at least a third-order Butterworth design with a cut-off frequency of 60 MHz. The filter should be designed to accommodate the maximum modulation index which would produce a frequency range of approximately 8 MHz to 56 MHz.

3.6.2.2 Demodulation and display

It was mentioned in Section 3.6.1 that a limiter-discriminator is the established demodulator circuit for FM-based modulation systems and hence a sample implementation is shown in Figure 3.20. In this circuit configuration a band pass filter with a range from 5 to 60 MHz is employed to prevent high frequency signal components from interfering with the demodulation process. The filtered signal is then amplified by an SL592 amplifier which produces complementary signal outputs, one of which is subsequently delayed relative to the other. The 10 dB gain elements in combination with the attenuators in these two paths match the power of the signals to the sensitivity of the relevant mixer input. The mixer output has a 6 dB attenuator in place which not only attenuates the signal but also acts as an impedance match to the video amplifier input. To eliminate interference from high frequency signal components generated at the mixer, a low pass filter should be inserted before the video amplifier. The low pass filter cut-off should be equal to the maximum baseband signal frequency and should be impedance-matched to the mixer output and video amplifier input.

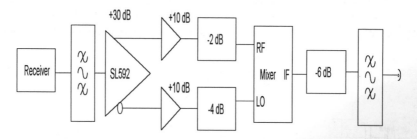

Figure 3.20 *Practical limiter-discriminator configuration*

Finally the video amplifier must transform the ±5 mV mixer output to a 0 to 700 mV signal at the display. An example of a configuration which can be used for this purpose is shown in Figure 3.21. The gain of the amplifier is set using the potentiometer and is variable from 6 to 112 but in practice the performance of the amplifier degrades at gains above 30 and hence two devices must be cascaded together to provide the required gain.

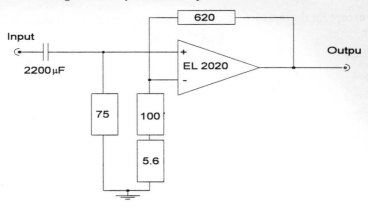

Figure 3.21 *Video amplifier configuration*

The action of a limiter-discriminator circuit is relatively straightforward in that a single frequency output from the receiver will produce a constant phase relation between the complementary signal outputs, thus giving a DC voltage at the mixer output which changes with frequency as shown in Figure 3.22. Changing the received signal frequency will change the DC voltage obtained as the fixed delay between the complementary signals will form a different proportion of the relative phase of the signals. This relative phase change will continue until a 180° change is attained.

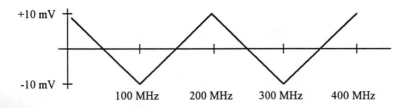

Figure 3.22 *Typical FM discriminator response*

Changing the relative path lengths of the RF and LO inputs will allow this frequency range to be tuned to the desired region. As in the case of the resistors for the VTO devices, a variable delay can be introduced to establish the exact region of interest and a fixed delay then substituted.

3.6.3 Square wave frequency modulation system implementation

The overall system configuration for an SWFM implementation is shown in Figure 3.23. This is somewhat similar to that proposed for an SCFM system (see Figure

3.13) except for the inclusion of a gain element and an attenuator in the modulating circuit.

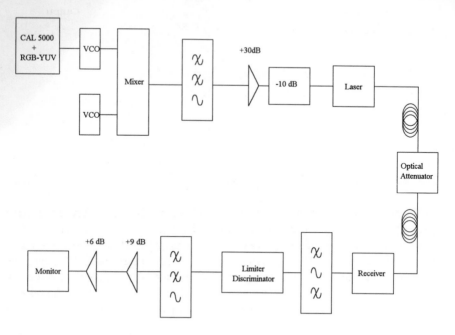

Figure 3.23 *SWFM system implementation*

3.6.3.1 FM signal shaping to form SWFM

The FM modulator described in Section 3.6.1 can be used as the source for the signal-shaping circuit which generates the SWFM signal. The output of the mixer (low pass filtered) is a sine wave varying from 8 to 56 MHz if the maximum modulation index is applied. The amplitude of the sine wave will be approximately 100 mV and, in the case of SCFM, a gain of approximately 10 should be provided resulting in a modulating signal being applied to the laser diode of approximately 1 V. In the SWFM implementation, a gain of 30 dB has to be applied to give the correct pulse shape, followed by attenuation of 10 dB implying the application of a 10 V signal to the laser diode. In this case, however, the 30 dB gain causes clipping of the amplified signal to the power rails thus limiting the signal applied to the laser diode to approximately 1 V.

The clipping circuit does not give an ideal square wave but does have the required frequency content. For a 25 MHz input frequency, the clipping circuit would yield a square wave with period 40 ns and rise and fall times of approximately 2 ns. To accomplish this performance, a monolithic microwave integrated circuit (MMIC) can be used, employing a circuit of the type shown in Figure 3.24. MMIC devices are generic in design and only require a change of bias resistor to allow for differences in operating current.

3.6.3.2 SWFM demodulation

The simplest method to demodulate SCFM is to utilise a limiter-discriminator. No modifications to the SCFM circuit are necessary as the frequency content of the band pass filtered SWFM signal at the receiver output should be identical to that of the SCFM system, assuming the same applied bandwidth. The demodulator therefore decodes the SWFM signal as if it were SCFM. An alternative method of demodulation is to convert the received signal into PFM, either by use of a monostable multivibrator, a D-type flip-flop or the use of exclusive OR gates [42], generating a baseband signal which can then be recovered utilising low pass filtering.

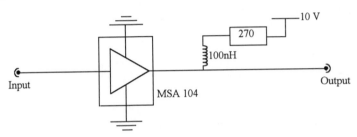

Figure 3.24 *Implementation of ECL amplifier with MMIC circuitry*

3.7 Evaluating the performance of modulation systems for TV transmission

Conventional television systems may be quantified using the traditional parameters of differential phase and differential gain. Direct SNR measurements can be made and these are useful to distinguish the relative performance of the analogue implementations. Another quantitative assessment method which can be employed for comparative performance evaluation is the measurement of the mean-square-error (MSE) [43]. This technique utilises frame stores to compare the original picture data with the transmitted signal and plots the difference in a two-dimensional array. A global figure of merit can then be calculated from this data. The two-dimensional data can also give important information on the cause of the distortion. If, for example, the MSE plot shows a large error after rapid changes in signal then an amplifier may be causing overshoot. Correct analysis does, however, depend strongly on the timing integrity of the frame stores utilised to make the comparison [44].

The problem of comparing the relative performance of modulation techniques has been partly addressed by standards bodies, in particular the CCIR. In order to carry out a meaningful evaluation of the comparative performance of a number of modulation methods it is necessary to set up a standard viewing system and assess the transmitted picture quality relative to an untransmitted signal. A procedure for carrying out this subjective performance comparison is detailed in a CCIR document [45]. Of the parameters mentioned above, SNR, DG and DP are the most informative, and SNR may be particularly appropriate when applied to analogue optical fibre transmission systems where the electro-optic and opto-electronic conversions may make significant contributions to the overall system performance. MSE may also be appropriate even though the relative SNR performance of the systems provides the major contribution to the MSE figure. Little additional performance information is therefore gained by carrying out this measurement. However, the CCIR subjective performance assessment can provide very meaningful data for analogue system testing.

The assessment of SNR performance is also detailed in a CCIR report [13] and the circuit configuration shown in Figure 3.25 incorporates these requirements. Essentially this testing procedure replaces the TV source with a 75 Ω load and the monitor with a switchable test circuit comprising either a video oscilloscope or an r.m.s. voltmeter with a 75 Ω load resistor and TV video weighting network. The test procedure is straightforward and requires a transmitted signal to be displayed on the video oscilloscope. It is then scaled so that the black level to white level voltage was 700 mV. Once the gain is correctly set then the video source should be replaced by the 75 Ω load and the circuit switched into its noise testing configuration for a reading of the noise voltage, v_n, to be taken. The SNR is then calculated using the following expression [17]:

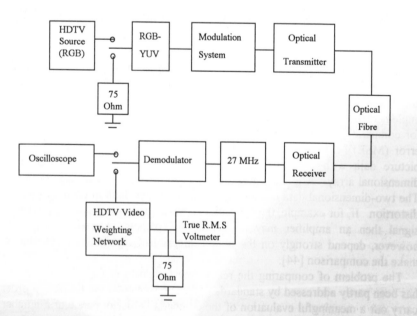

Figure 3.24 *Signal-to-noise ratio test circuit*

$$\text{SNR}_{PFM} = 20 \log_{10}\left(\frac{700}{v_n}\right)$$

$$(3.58)$$

Adjustment of the received optical power (for example with an optical attenuator) allows a range of receiver sensitivities and corresponding SNRs to be evaluated and plotted to enable characterisation of the overall system response. It should be noted, however, that it is essential that the voltmeter used in this evaluation should have sufficient bandwidth to capture all of the noise content of the transmitted signal.

Direct measurement of the optical signal from the output of the optical attenuator is possible using an electro-optic converter and this allows the optical modulation depth to be evaluated. This value can then be used, along with the measured SNR values, to draw a comparison between theoretical expectations and practical implementations. The SNR data is also important to define the conditions for the CCIR subjective picture quality testing [45]. In order to simulate the practical limitations of a fixed optical installation the assessments should be performed with a fixed receiver sensitivity. Hence using this configuration any improvement in transmitted picture quality is due to an improved modulation efficiency, resulting in increased SNR. The alternative is to carry out the CCIR comparisons with the analogue SNR maintained the same for each system. However, as each system then exhibits the same proportion of noise content, this provides a test to indicate how obtrusive the noise pattern for a particular modulation system is relative to that of the other systems, rather than a measure of their relative modulation efficiencies in a practical application. In summary, however, it is almost certain that the subjective testing strategy will remain the major performance assessment of the overall transmission quality

3.8 References

1   Senior, J. M.: *'Optical Fiber Communications : Principles and Practice'*, Second *Edition*, (Prentice-Hall International, New York, 1992).

2   Olshansky, R., Lanzisera, V. A., and Hill, P. M.: 'Subcarrier multiplexed lightwave systems for broadband distribution', *J. Lightwave Technol*, 1989, **7**, pp.1329-1341

3   Darcie, T. E.: 'Subcarrier multiplexing for lightwave networks and video distribution systems', *IEEE J. Selected Areas in Commun.*, 1990, **8**, pp.1240-1248

4   Yoneda E., Suto, K., Kikushima, K. and Yoshinaga, H.: 'Fully engineered multi-channel FM-SCM video distribution systems', *J. Lightwave Technol.*, 1994, **12**, pp.362-368

5   Forrest, S. R.: 'The sensitivity of photoconductor receivers for long-wavelength optical communications', *J. Lightwave Technol.*, 1985, **LT-3**, pp.347-360

6   Kasper, B. L.: 'Receiver Design', in *Optical Fibre Telecommunications II*, Ed. Miller, S. E. and Kamonow, I. P., (Academic Press Inc. Ltd, London, 1988)

7   Muoi, T. V.: 'Receiver design for high-speed optical-fibre systems', *J. Lightwave Technol.*, 1985, **LT-2**, pp.243-267

8   Goell, J. E.: 'An optical repeater with high impedance input amplifier', *Bell Syst. Technol. J.*, 1974, **53**, pp.629-643

9   Schwartz, M.: *'Information Transmission, Modulation and Noise, Fourth edition'*, (McGraw-Hill, London, 1990)

10 Ogawa, K.: 'Considerations for optical receiver design', *IEEE J. Selected Areas in Commun.*, 1983, **SAC-1**, pp.524-532

11 Ogawa, K.: 'Noise caused by GaAs MESFETs in optical receivers', *Bell Syst. Technol. J.*, 1981, **60**, pp.650-652

12 Brain, M. and Lee, T. P.: 'Optical receivers for lightwave communication systems', *J. Lightwave Technol.*, 1985, **LT-3**, pp. 1281-1300

13 Ohkawa, N.: '20 GHz low noise HEMT preamplifier for optical receivers', *European Conference on Optical Communications*, Brighton, England, September 1988, pp. 404-407

14 O'Mahony, M. J.: 'The analysis and optimisation of a PIN photodiode GaAs FET cascode optical receiver', *British Telecom Technol. J.*, 1983, **1**, pp.38-42

15 Mendis, F. V. C. and Rosher, P. A.: 'CNR requirements for subcarrier multiplexed multichannel video FM transmission in optical fibre', *Electron. Lett.*, 1989, **25**, pp.72-74

16 Kashima, N.: *Optical Transmission for the Subcarrier Loop* (Artech House, Norwood, Massachusets, USA, 1993)

17 CCIR Report 637-2, 'Signal-to-noise-ratio in television', *XVth Plenary Assembly*, Geneva, 1982, **XII**, pp.41-45

18 Kakimoto, S., Ohkura, Y., Takemoto, A., Yoshida, N., Namazaki, H., Susaki, W. and Shibayama, K.: 'Extremely low threshold 1.3mm DFB-PPIBH laser diode applied to 1 Gbit/sec zero bias RZ modulation', *European Conference on Optical Communications*, Brighton, 1988, pp.337-340

19 Chiddix, J. A.: 'AM transmission of video through fiber', *16th International Television Symposium, CATV sessions*, Montreux, 1989, pp. 131-137

20 Taub, H. and Schilling, D. L.: *Principles of Communication Systems* (McGraw-Hill, London, 1986)

21 Luvison, A.: 'Topics in optical fibre communication theory', *Optical Fiber Communications* by Technical Staff of CSELT, pp. 647-721 (McGraw-Hill, London, 1981)

22 Viterbi, A. J.: *Principles of Coherent Communications* (McGraw-Hill, London, 1966)

23 Prisco, J. J.: 'Fiber optic regional area networks in New York and Dallas', *IEEE J. Selected Areas in Communications*, 1986, **SAC-4**, pp. 750-757

24 Way, W. I.: 'Subcarrier multiplexed lightwave system design considerations for subscriber loop applications', *J. Lightwave Technol.*, 1989, **7**, pp. 1806-1818

25 Dobbie, W. H.: 'Narrow-band FM TV transmission', *International Broadcasting Convention 1988*, Brighton, England, pp. 300-303

26 Kibe, S. V.: 'Zero crossing analysis of FM threshold', *IEEE Trans. Commun.*, 1982, **COM-30**, pp. 1249-1254

27 Suh, S. H.: 'Pulse width modulation for analog fiber-optic communication', *J. Lightwave Technol.*, 1987, **LT-5**, pp. 102-112

28   Martin, J. D. and Hausien, H. H.: 'PPM versus PCM for optical local-area networks', *IEE Proceedings-I*, 1992, **139**, pp. 241-250

29   Wilson, B. and Ghassemlooy, Z.: 'Differential phase error in pulse-width-modulated TV transmission systems', *Electron. Lett.*, 1987, **23**, pp.1133-1134

30   Marougi, S. D. and Sayhood, K. H.: 'Noise performance of pulse interval modulation system', *Int. J. Electron.*, 1983, **55**, pp. 603-614

31   Marougi, S. D. and Sayhood, K.: 'Signal to noise performance of the pulse interval and width modulation system', *Electron. Lett.*, 1983, **19**, pp. 528-530

32   Webb, R. P.: 'Output noise spectrum from demodulator in an optical PFM system', *Electron. Lett.*, 1982, **18**, pp. 634-636

33   Lathi, B. P.: *Modern Digital and Analog Communication Systems* (Holt-Rinehart and Winston, New York, 1983)

34   Brace, D. J.and Heatley, D. J.: 'The application of pulse modulation schemes for wideband distribution to customers' premises', *European Conference on Optical Communications*, York, 1980

35   Chen, C. C. and Gardner, C. S.: 'Performance of PLL synchronized optical PPM communication systems', *IEEE Trans. Commun.*, 1986, **COM-34**, pp. 988-994

36   Sedra, A. S. and Smith, K. C. *'Microelectronic Circuits'* (Holt, Reinhart and Winston, London, 1982)

37   Timmerman, C. C.: 'Signal-to-noise-ratio of a video signal transmitted by a fiber optic system using pulse-frequency modulation', *IEEE Trans. Broadcast.*, 1977, **BC-23**, pp.12-16

38   Drukarev, A. I.: 'Noise perormance and SNR threshold in PFM', *IEEE Trans. Commun.*, 1982, **COM-30**, pp. 708-711

39   Hillerich, B., Rode, M. and Weidel, E.: 'Grating multiplexer with wide passband', *European Conference on Optical Communications*, Stuttgart, 1984, pp. 168-269

40   Olshansky, R. and Lanzisera, V. A.: '60-channel FM video subcarrier multiplexed optical communication system', *Electron. Lett.*, 1987, **23**, pp. 1196-1197

41   Way, W. I., Zah, C. E., Menocal, S. G., Caneau, C., Favire, F., Shokoohi, F. K., Lee, T. P. and Cheung, N. K.: '90-channel video transmission to 2048 terminals using two inline travelling wave laser amplifiers in a 1300nm subcarrier multiplexed optical system', *European Conference on Optical Communications*, Brighton, England, September   1988, pp. 37-40

42   Wilson, B., Ghassemlooy, Z. and Lu, C.: 'Optical fibre transmission of high definition television signals using square wave frequency modulation', *Third Bangor Symposium on Optical Communications*, Bangor, Wales, 1991, pp. 258-261

43   Weaver, L. E. *'Television Measurement Techniques',* IEE Monograph Series 9, (Peter Peregrinus Publishing, London, 1971)

44   Baraclough, J. N. and Main, A. B.: 'Experiences with digital video components', *International Broadcasting Convention 1986*, Brighton, England, 1986, pp. 280-283

45   CCIR Recommendation 500-3, 'Method for the subjective assessment of the quality of television pictures', *Recommendations and reports of the CCIR*, 1986, No. 11, Part 1

Appendix 3.1

Typical component values for a high quality pin-FET optical receiver (for use with Equations 3.13, 3.14, 3.17, 3.19 and 3.20). The variables are listed in the order they appear in Equation 3.8 first, then Equation 3.9.

Variable	Typical value	Variable	Typical value
I_d	20 nA	C_J	0.9 pF
K	1.38 E-23 J K^{-1}	C_T	1.2 pF
T	293 K	R_3	545 Ω
R_B	300 KΩ	R_4	100 Ω
I_G	20 nA	F_N	3 dB
Γ	0.7	R_A	50 Ω
g_M	30 mS	f_B	50 MHz
R_S	10 Ω	f_L	10 kHz
C_C	2.2 pF - dependent on frequency compensation required		

Chapter 4

Pulse time modulation techniques for optical fibre communications

B. Wilson and Z. Ghassemlooy

4.1 Introduction

Optical fibre point-to-point links and networks are being used to provide broadband telecommunication services which utilise multiplexes of video, data and voice channels. The choice of modulation format on the optical carrier is therefore a principal factor in realising high-performance bandwidth-efficient transmission of high-quality analogue sourced signals at an acceptable cost. Even though the bandwidths offered by new fibre installations are very high, they are still limited by fibre dispersion and carrier incoherence over practical distances.

Analogue modulation schemes, in which the optical source is modulated in a continuous manner, are both simple and bandwidth-efficient, but often cannot deliver the required signal-to-noise ratio over anything other than short distances [1,2]. In addition, these schemes suffer to a certain extent from non-linearity of the optical channel and associated circuitry, severely limiting the quality of the received information through intermodulation and crosstalk. In contrast, digital schemes such as pulse code modulation (PCM) have been demonstrated to be substantially immune to channel non-linearity and are capable of producing the required signal-to-noise ratio. However, digital systems are significantly more complex and costly than analogue systems, largely due to their coding circuitry and intrinsically large bandwidth overhead. This overhead can be reduced towards that of analogue schemes by employing code compression, but then complexity increases significantly and so overall cost remains high.

Pulse time modulation (PTM) represents an alternative approach that occupies an intermediate position between analogue and digital modulation formats [3]. Modulation is simple, requiring no digital coding, while the pulse format of the modulated carrier renders the scheme immune to channel non-linearity and allows routing through digital switching nodes and logic circuits in a network.

Moreover, PTM is unique in its ability to trade-off performance with bandwidth overhead, which is a particularly exploitable feature in fibre systems [3,4]. For example, certain short distance applications, such as local area networks (LANs), may use multimode or monomode fibre with a dispersive optical carrier, all of which impose a significant bandwidth limitation in the optical channel. In these applications low-speed optical sources such as light-emitting diodes (LEDs) can be used with PTM, while still achieving the required signal-to-noise ratio. In contrast, the available bandwidth on long-distance terrestrial and undersea routes may be many orders of magnitude broader where optical amplification and soliton techniques are employed. This additional bandwidth can be readily exploited by PTM to improve performance and signal-to-noise ratio. The ability to exchange signal-to-noise performance against bandwidth extension is a property unique to PTM modulation techniques and is of increasing importance in high-speed networks. PTM techniques are of particular interest where short pulses, such as solitons, may be employed, since further forms of coding can yield further significant improvements over PCM.

Adoption of PTM techniques has certain beneficial consequences from the standpoint of optoelectronic sub-system specifications. Since PTM deals exclusively with a pulse format there are no concerns over LED or laser diode linearity, as would be the case with direct intensity modulation or subcarrier multiplexing techniques. In addition, for narrow pulse PTM formats the peak optical output may be maintained at a high level to ensure good noise performance at the receiver without compromising device lifetime through elevated mean transmitter power levels. An optical transmitter for a PTM system may therefore be chosen primarily for its maximum peak power level in order to maximise transmission distance and signal-to-noise performance with little regard for device linearity. LEDs are attractive in both cost-effectiveness and circuit simplicity, but inevitably suffer from restricted transmission distance through fibre dispersion limitations at shorter wavelengths, whereas injection laser diodes (ILDs) offer higher launch powers and narrower linewidths for longer wavelength high capacity systems.

4.2 Pulse time modulation family

The basic framework of research into PTM techniques was laid down around 50 years ago and reported in the late 1940s [5-10], but it is only recently that a revival of interest has been experienced with the development of fibre transmission systems [3].

In all pulse time modulation methods one of a range of time dependent features of a pulsed carrier is used to convey information in preference to the carrier amplitude, as outlined in the table illustrated in Figure 4.1.

PTM type	Variable
PPM	Position
PWM	Width (duration)
PIM	Interval (space)
PIWM	Interval and width
PFM	Frequency
SWFM	Frequency

Figure 4.1 *PTM family*

In pulse width modulation (PWM), sometimes referred to as pulse duration modulation (PDM), the width of the pulsed carrier within a predetermined time frame is changed according to the sampled value of the modulating signal, Figure 4.2(a). Pulse position modulation (PPM) may be considered as differentiated PWM and carries information by virtue of the continuously variable position of a narrow pulse within a fixed time frame, as in Figure 4.2(b). Digital PPM employs discrete time slots to represent expanded PCM, Figure 4.2(c)

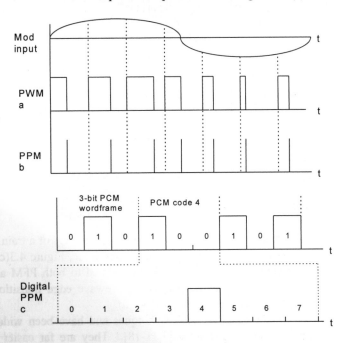

Figure 4.2 *Isochronous PTM techniques*

As its name suggests, in pulse interval modulation (PIM) the variable intervals between adjacent narrow pulses are determined by the amplitude of the input signal, Figure 4.3(a). Pulse interval and width modulation (PIWM) is derived directly from PIM to produce a waveform in which both mark and space convey information in alternating sequence, Figure 4.3(b). Each successive time frame in both PIM and PIWM commences immediately after the previous pulse, unlike PWM and PPM which have fixed time frames allocated for sampling. Digital PIM employs discrete time slots in a manner comparable to DPPM and exhibits relaxed synchronisation criteria by virtue of its anisochronous nature.

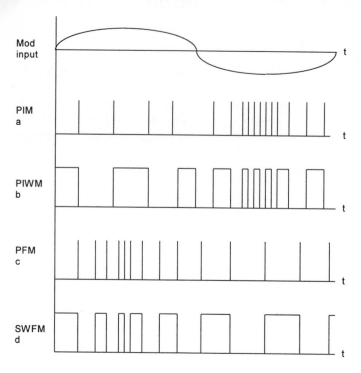

Figure 4.3 *Anisochronous PTM*

In pulse frequency modulation (PFM) the instantaneous frequency of a train of narrow pulses is determined by the modulating signal amplitude, Figure 4.3(c). Squarewave frequency modulation (SWFM) is closely related to both PFM and analogue FM, consisting essentially of a series of squarewave edge transitions occurring at the zero crossing points of FM, Figure 4.3(d).

PWM and PPM are both long-established techniques and have been widely adapted for use in optical fibre applications [3,11-18]. They are far easier to multiplex in the time domain because of their fixed frame timing intervals and require only a moderately simple demultiplexor at the receiving end. PFM has been used extensively for optical fibre transmission of video and broadcast quality TV signals [19-24], with SWFM being employed for fibre transmission of HDTV and other wideband instrumentation signals [3,25-30]. Comparatively little work however has been published on PIM and PIWM applied to wideband fibre transmission [3,20, 31-34]. Recent work on digital PPM has been aimed primarily at long-haul fibre and free-space applications [35-37]. All PTM methods are of course equally amenable to wavelength division multiplexing.

4.3 Analysis

4.3.1 PTM modulation spectrum

All PTM techniques produce modulation spectra that share a common set of features. In each case modulation gives rise to a diminishing set of sidetones centred around the carrier (sampling) frequency and its harmonics, separated in frequency by an amount equal to the signal frequency, as illustrated in Figure 4.4. The number and strength of the sidetones (the sidetone profile) is a characteristic unique to each PTM technique and provides a signature by which it may be recognised. In addition, a baseband component is also present for some PTM methods, along with harmonics, depending upon the form of sampling employed .

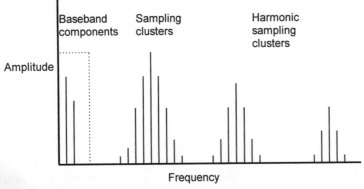

Figure 4.4 *Typical PTM modulation spectrum*

Either natural or uniform sampling of the input signal may be adopted for PTM. Naturally sampled modulators operate directly on the input signal such that the precise sampling instants are variable and determined by the location of the modulated pulse edges. For uniform sampling the input signal is routed via a sample-and-hold circuit which produces flat-topped amplitude modulated pulses, so that the PTM modulator is operating on uniformly spaced and stored input samples.

Demodulation for natural sampling is effected, after amplitude threshold detection in a PTM receiver, by converting the particular PTM waveform under consideration into a form with a baseband component and filtering out the carrier and sidetone components by a low-pass filter. When uniform sampling is employed a sample-and-hold circuit is included to recreate an amplitude modulated pulse form, followed again by a low-pass filter. The choice between natural and uniform sampling is essentially one of performance-cost trade-off, since uniform sampling is in principle capable of complete signal recovery without distortion, but with the additional expense of a premium specification sample-and-hold unit for high-speed signals.

All PTM techniques display a noise threshold effect below which signal pulses become indistinguishable from noise pulses. Above threshold, noise occurring during the leading or trailing edges of received PTM pulses manifests itself as timing jitter in the regenerated pulses, and hence as amplitude noise after demodulation. The slope of the received PTM pulse edges determines the period in which noise is able to influence the decoding timing decisions, and therefore the quality of the recovered signal. This mechanism results in demodulated noise power being inversely proportional to the square of the ratio of transmission channel bandwidth to carrier frequency [38] and is responsible for the very useful attribute in PTM of being able to trade-off channel bandwidth against signal-to-noise performance.

The approach widely adopted for investigations into PTM modulation spectra is based on a Fourier series expansion of a pulse train modulated by a single sinewave frequency [38-40], rather than assuming a random input signal and employing the autocorrelation of the process to obtain the spectral density of the modulated waveform. In a broad range of transmission applications, such as TV signals, this approach is justified since the analogue signals will display a dominant subcarrier frequency.

4.3.2 Classification of PTM techniques

A general method of classifying and categorising PTM modulation techniques has only recently been proposed, based on the spectral descriptions outlined in the following sections and by considering the behaviour of the sampling waveform's fundamental spectral component under modulation conditions [3].

For both PWM and PPM this spectral component remains constant in frequency and never becomes zero under any conditions of modulation, since the internal ramps employed within the modulators are permitted to run their full course without influence from the modulating signal. For this reason the term "isochronous" has been adopted for describing PWM and PPM by virtue of the equal sampling intervals available within their modulation techniques. The duration of the sampling period for both PWM and PPM is determined solely by the choice of carrier (sampling) frequency and is not affected in any manner by the amplitude or frequency of the modulating signal.

However, for the other forms of PTM the equivalent spectral component is seen to either vanish for certain values of modulation index (PFM and SWFM) or change in both amplitude and frequency (PIM and PIWM) as a result of the nature of the modulation process. In both cases the effective sampling period utilised within the respective modulators is modified dramatically by the presence of the modulating signal. The classification term "anisochronous" is therefore most appropriate in these cases to reflect the unequal sampling periods employed under modulation conditions.

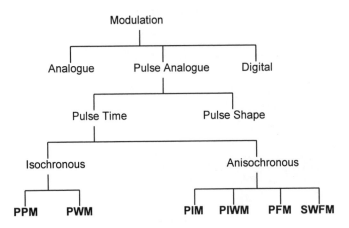

Figure 4.5 *PTM classification system*

The proposal of a PTM classification system based on the concept of (an)isochronous sampling, as illustrated in Figure 4.5, is one which fits naturally with both theoretical and practical considerations. Inspection of spectral predictions from the various modulation formulae will indicate the correct classification category for all existing and any new techniques, while practical spectral measurements will similarly yield an immediate answer by direct observation. Classification in this manner aids modulator testing by predicting suitable calibration features, but has no overt influence on demodulator design, since, in general, knowledge of carrier frequency behaviour is not required for successful signal recovery.

4.3.3 Pulse width modulation

In pulse width modulation (PWM), the width of the pulsed carrier is changed according to the sampled value of the modulating signal, as previously illustrated in Figure 4.2(a). Leading, trailing or double edge modulated PWM may be generated by comparison of the input signal with a linear ramp waveform, or triangular in the case of double edge modulation. For naturally sampled PWM this comparison is performed directly at a comparator, whereas in uniformly sampled PWM the input signal is routed first through a sample-and-hold circuit so that the input samples are taken at evenly spaced intervals rather than at varying intervals dependent on the signal amplitude, as in Figure 4.6(a). For all styles of PWM (and PPM) a modulation index M ($0<M<1$) may be defined such that maximum modulation occurs when the peak-to-peak amplitude of the input signal is equal to the ramp amplitude, equivalent to a symmetrical peak-to-peak displacement equal to the sampling period for the PWM variable edge or PPM pulse.

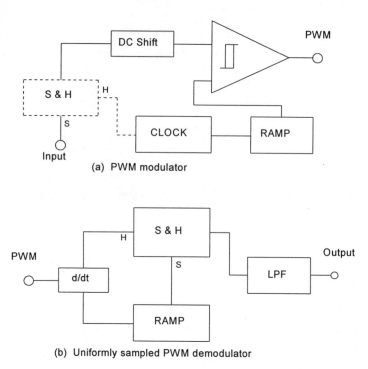

(a) PWM modulator

(b) Uniformly sampled PWM demodulator

Figure 4.6 *PWM modulator and demodulator*

A comprehensive treatment of the Fourier structure applicable to PWM with a single input tone has been given by Black [38], Bennett [39], Stuart [40] and Barton [41], who have shown that a trailing edge modulated, naturally sampled PWM waveform with a unity unmodulated mark/space ratio may be expressed as:

$$u(t) = \frac{1}{2} - \frac{M}{2}\sin\omega_m t + \sum_{n=1}^{\infty}\frac{\sin(n\omega_c t)}{n\pi} - \sum_{n=1}^{\infty}\frac{J_0(n\pi M)}{n\pi}\sin(n\omega_c t - n\pi)$$

$$- \sum_{n=1}^{\infty}\sum_{k=\pm1}^{\pm\infty}\frac{J_k(n\pi M)}{n\pi}\sin[(n\omega_c + k\omega_m)t - n\pi]$$

(4.1)

where M is the modulation index ($0<M<1$) and ω_m and ω_c are the modulating signal and carrier frequencies respectively. $J_k(x)$ is a Bessel function of the first kind, order k.

The second term in Equation 4.1 represents the baseband component of ω_m, while the combined effect of the third and fourth terms is to produce components at the carrier frequency ω_c and its harmonics. In the unmodulated case ($M=0$), only odd harmonics are created, whereas the even harmonics are introduced as M increases. Term 5, involving a double summation, represents the characteristic PTM series of diminishing sidetones set around the carrier frequency and all its harmonics and separated by a frequency equal to that of the input signal ω_m, Figure 4.7(a). It can be seen from Equation 4.1 that the frequency of the sampling (carrier) component is not affected by the value of the modulation index M and that its magnitude never falls to zero for any level of modulation. For this reason PWM can be classified as an isochronous PTM technique.

When uniform sampling is employed the resulting modulation spectrum is very similar [39], except that the baseband now contains in addition a diminishing harmonic series of the modulating signal, influenced not just by the modulation index M, but also by the ratio between the carrier and input signal frequencies.

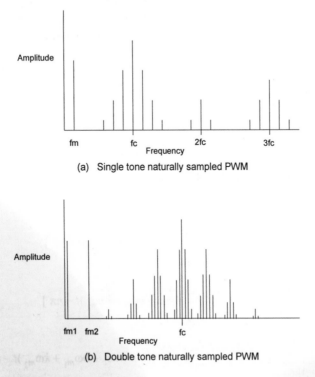

(a) Single tone naturally sampled PWM

(b) Double tone naturally sampled PWM

Figure 4.7 *PWM modulation spectra*

Demodulation in naturally sampled PWM is by threshold detection and a simple low-pass filter to recover the baseband component directly, whereas for uniformly sampled PWM, with its baseband harmonics, conversion into pulse amplitude modulation (PAM) is necessary before low-pass filtering, Figure 4.6(b). The additional complexity of uniformly sampled PWM is offset by the higher maximum modulation index that may be employed and hence the improved maximum signal-to-noise ratio available.

Demodulation of PWM (and PPM) may also be effected by an iterative reconstruction technique implemented via digital signal processing (DSP) with a resulting improvement in signal-to-noise ratio at low sampling ratios. However, due to the severe processing overheads involved this approach is only of significance for signals with bandwidths in the kilohertz region [42].

The basic modulation formula employed to describe and calculate the structure of the PWM spectral modulation components has proved durable by virtue of its accuracy in predicting the amplitude and frequency of all the sidetone components centred around the sampling frequency and its harmonics. However, it is unable to predict the complex PWM sidetone structure resulting from two or more independent input signals such as may occur with a complex modulating signal employing subcarriers. As PWM is a nonlinear process the principle of superposition does not apply and so when two or more independent input signals are considered it is not possible to add their individual Fourier series to give the composite spectrum [43]. Viewing the modulated squarewave as a waveform composed of positive and negative staircases, rather than the original double Fourier series adopted by Stuart [40], the authors have recently derived a spectral prediction formula for two-tone input PWM as [44]:

$$u(t) = \frac{1}{2} - \frac{M_1}{2}\sin\omega_1 t + \frac{M_2}{2}\sin\omega_2 t + \sum_{n=1}^{\infty}\frac{\sin(n\omega_c t)}{n\pi}$$

$$-\sum_{n=1}^{\infty}\frac{J_0(n\pi M_1)J_0(n\pi M_2)}{n\pi}\sin(n\omega_c t - n\pi)$$

$$-\sum_{n=1}^{\infty}\sum_{h=\pm1}^{\pm\infty}\frac{J_0(n\pi M_2)J_h(n\pi M_1)}{n\pi}\sin[(n\omega_c + h\omega_{m_1})t - n\pi] \qquad (4.2)$$

$$-\sum_{n=1}^{\infty}\sum_{k=\pm1}^{\pm\infty}\frac{J_0(n\pi M_1)J_k(n\pi M_2)}{n\pi}\sin[(n\omega_c + k\omega_{m_2})t - n\pi]$$

$$-\sum_{n=1}^{\infty}\sum_{h=\pm1}^{\pm\infty}\sum_{k=\pm1}^{\pm\infty}\frac{J_h(n\pi M_1)J_k(n\pi M_2)}{n\pi}\sin[(n\omega_c + h\omega_{m_1} + k\omega_{m_2})t - n\pi]$$

Equation 4.2 represents a naturally sampled PWM waveform employing a square carrier wave of frequency ω_c modulated by two independent frequencies ω_{m1} and ω_{m2} with modulation indices M_1 and M_2 respectively. The novelty of the equation resides in the triple summation of term 8, whereby a series of diminishing sub-sidetones of spacing ω_{m2} set around all the ω_{m1} diminishing sidetones is generated. Figure 4.7(b) illustrates the character of this sub-sidetone structure set around all the primary sidetones along with the sidetones around the carrier fundamental. If either ω_{m1} or ω_{m2} is removed the spectral structure reverts to the usual single tone modulation pattern predicted by the original formula, since in either case $J_0(0)=1$ in the appropriate term 6 or 7, along with $J_{\neq0}(0)=0$ for term 8.

An enhanced ability to predict these PWM sub-sidetone structures is of importance in applications involving subcarrier modulation techniques, or with TV and video signals, where strong simultaneous input frequency components are present in the form of colour and sound subcarriers.

When employing PWM, or indeed any of the PTM techniques, for transmission of TV and video it is possible to generate low-level "beats" in the demodulated picture because of interactions between the carrier (sampling) frequency and the colour subcarrier. To avoid this problem it is necessary to adopt an integral ratio between the PTM carrier frequency and the colour subcarrier frequency. A value of 3:1 is the lowest practicable ratio because of the nature of PAL and NTSC broadcast TV signal formats and sampling requirements. A higher ratio will result in improved picture quality, since it has a direct influence on reconstructed differential phase error acting via sidetone coincidence [45].

4.3.4 Pulse position modulation

In PWM, the wasted portion of transmitted power that conveys no information depends on the maximum level of modulation that may be employed. Pulse position modulation (PPM) is the result when the wasted element is subtracted from PWM. This power saving represents the fundamental advantage of PPM over PWM. PPM may be considered as differentiated PWM and carries information by virtue of the continuously variable position of a narrow pulse within a fixed time frame, as in Figure 4.2(b). Since the pulsed carrier in this format may be made very narrow it is well suited for use with injection laser diodes (ILDs). A typical naturally sampled PPM modulator is very similar to a PWM modulator and consists simply of a comparator detecting equivalence between the input signal and a linear ramp, followed by a monostable or other pulse-generating circuit. Since the main feature of PPM is its power efficiency, no additional pulses are transmitted for frame timing purposes, these being reconstructed within the demodulator.

Employing similar notation as in previous sections, the naturally sampled PPM modulation spectrum may be represented by [40]:

$$f(t) = \frac{A\omega_c \tau}{2\pi} + AM\cos(\omega_m t)\sin(\omega_m \tau / 2)$$

$$+ \frac{2A}{\pi} \sum_{k=-\infty}^{\infty} \sum_{n=1}^{\infty} J_k(n\pi M) \frac{\sin[(n\omega_c + k\omega_m)\tau / 2]}{k} \cos[(n\omega_c + k\omega_m)t] \qquad (4.3)$$

where τ represents the width of PPM pulses of amplitude A.

Spectral components are generated at the sampling frequency and all its harmonics along with diminishing groups of sidetones separated by the modulating frequency. Modulation sidetones centred around the sampling frequency and all its harmonics are influenced both by the carrier pulse width and the modulating frequency. If the pulse duration is increased the overall sampling frequency harmonic profile is modified by a sinc(x) function and eventually becomes similar to that of PWM. This has little effect in practice, since, in general, only low-order carrier harmonics are retained in most transmission applications.

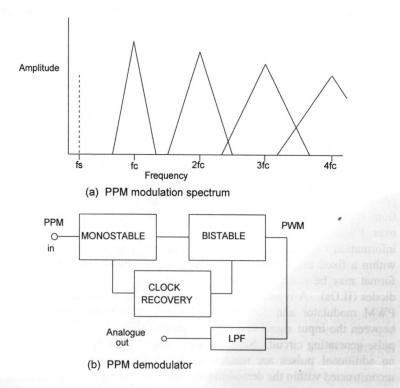

(a) PPM modulation spectrum

(b) PPM demodulator

Figure 4.8 *PPM modulation spectra and demodulator*

The PPM modulation spectrum also contains a baseband compone
of a differentiated version of the modulating signal, whose amplitude
on both pulse width and input frequency. Demodulation in its simplest ... is by
integration and low-pass filtering after threshold detection, but more usually
consists of converting the PPM pulse stream to PWM using bistable and
recovered clock pulses and then low-pass filtering to recover the original signal
[46], Figure 4.8(b). Correct reconstruction of clock pulses is usually arranged by
a simple phase-locked loop technique arranged to servo the DC component of the
reconstructed signal to zero [47]. If a DC-coupled transmission capability is
desired, then frame timing pulses must be included in the PPM pulse stream.

Digital PPM is a relatively new method in which the analogue message is first
digitally encoded as in a PCM system. This digital signal is then further encoded
by replacing packets of typically up to 16 bits with a single PPM pulse whose
position allocation within the frame interval is determined uniquely by the code
sequence of the associated packet [35,36], Figure 4.2(c). Improved detection
sensitivity can then be traded against clock-rate. This modulation scheme is
aimed primarily at long-haul fibre backbones and deep-space communications
where it is desirable to trade transmission bandwidth with repeater spacing, and
where the cost of coding complexity can be readily absorbed.

4.3.5 Pulse interval modulation

As its name suggests, in pulse interval modulation (PIM) the interval between
adjacent pulses is determined by the amplitude of the input signal. The format
falls into the anisochronous category since the duration of each sampling episode
is determined by the input modulating signal [48].

Figure 4.9(a) illustrates one possible implementation of a PIM modulator where a
comparator and ramp generator are connected as a feedback loop to reset the ramp
when equivalence is detected between the ramp and the DC-shifted input signal.
In the absence of an input signal the PIM output is a series of uniformly spaced
narrow pulses at a free-running frequency determined by both the ramp slope and
the unmodulated DC level. The addition of a DC level to the input signal is
necessary to ensure that there is always sufficient headroom to accommodate the
full dynamic range and to enable the ramp to sample the most negative input
swings. A sample-and-hold unit is inserted in the input lead for uniform
sampling, triggered from the comparator output. This is omitted for natural
sampling. A modulation index M ($0<M<1$) may be defined in this context as the
peak-to-peak modulating signal swing divided by twice the DC level. Operation at
very high levels of modulation index should normally be avoided with PIM since
the negative swing of the shifted modulating signal approaches zero and hence the
instantaneous frequency of the resulting PIM pulse train increases rapidly.

Following the same general approach as Fyath [49], based on earlier derivations [50,51], but adopting an expansion based on multiple angles, we may present the PIM modulation spectrum as:

$$p(t) = \left(\frac{\omega_0}{2\pi} \sum_{p=0}^{\infty} M^P \cos^P \omega_m t\right) \tag{4.4}$$

$$\times \left\{1 + 2 \sum_{k=\pm 1} \sum_{n_1=-\infty}^{\infty} \sum_{n_2=-\infty}^{\infty} J_{n_1}(kB_1) J_{n_2}(kB_2) \cos[(kA_0\omega_0 + (n_1 + 2n_2)\omega_m)t]\right\}$$

where

$$A_p = \sum_{r=0}^{\infty} \left(\frac{M}{2}\right)^{2r} \frac{(p+2r)!}{(p+r)!r!} \qquad (p = 0,1,2)$$

and

$$B_p = \frac{2\omega_0}{p\omega_m}(M/2)^P A_p \qquad (p = 1,2)$$

where the frequency of the unmodulated unity-strength impulse stream is ω_0, with an input signal frequency of ω_m.

The sidetone spectral profiles around the modulated sampling frequency and its-harmonics resulting from Equation 4.4 are somewhat asymmetrical and a strong function of the modulation index, Figure 4.9(b). There is also a series of baseband harmonics generated in addition to the original baseband component. In contrast to PPM, the magnitude of the baseband components is not influenced by the input frequency. As the depth of modulation is increased the average sampling frequency increases from its free-running value.

PIM demodulation is effected for low values of modulation index, typically less than 10%, by simply low-pass filtering to regain the original modulating signal, along with a diminishing series of harmonic distortion components. A more general approach, applicable to the full modulation range, uses the PIM pulses to reset and initiate a ramp whose maximum values constitute sampled points on the reconstructed modulating waveform, Figure 4.9(c). Final filtering takes place either by a low-pass filter, or a combination of sample-and-hold followed by a low-pass filter if uniform sampling has been adopted in the modulator. The noise performance of PIM is marginally superior to PPM under similar conditions [31,52], but displays a frequency squared term in its noise power spectral density which can be problematic in certain applications[53].

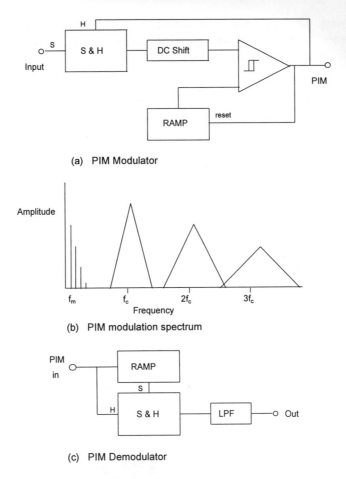

(a)   PIM Modulator

(b)   PIM modulation spectrum

(c)   PIM Demodulator

Figure 4.9 *PIM spectra and demodulator*

4.3.6 Pulse interval and width modulation

Pulse interval and width modulation (PIWM) is derived directly from its counterpart PIM, by passing PIM pulses through a bistable circuit to produce a waveform in which both mark and space convey information. Both natural and uniform sampling may be employed in a modulator design identical to the PIM case (Figure 4.9(a)) except for the addition of a bistable element at the output. As with PIM the anisochronous nature of PIWM arises because the modulator ramp is reset at a point in time determined by the instantaneous value of the input signal and not by a predetermined interval controlled by the choice of carrier (sampling) frequency.

Employing the same general approach as Fyath [49] for PIM, but by adopting an expansion based on multiple angles, the authors have recently obtained a general PIWM spectral expression given by [34,54]:

$$g(t) = \frac{V}{2}\left\{ 1 + \sum_{k=1}^{\infty}\sum_{n_1=-\infty}^{\infty}\sum_{n_2=-\infty}^{\infty} \frac{\sin(k\pi/2)}{(k\pi/2)} J_{n_1}(kB_1) J_{n_2}(kB_2) \right.$$

(4.5)

$$\left. \times \cos[(kA_0\omega_0 + (n_1+2n_2)\omega_m)t] \right\}$$

where

$$A_p = \sum_{r=0}^{\infty}\left(\frac{M}{2}\right)^{2r} \frac{(p+2r)!}{(p+r)!r!} \qquad (p=1,2)$$

and

$$B_p = \frac{2\omega_0}{p\omega_m}(M/2)^p A_p$$

M is the modulation index ($0<M<1$), V is the amplitude of the PIWM waveform, A_0 is calculated from A_p, with $p=0$.

Equation 4.5 is a close approximation based on a restricted expansion in which $p=1,2$ only, since both B_p and $J_n(kB_p)$ diminish very rapidly for higher values of p. The PIWM spectral profile resulting from Equation 4.5 is asymmetrical and contains no baseband component, in contrast to PIM. Strong spectral components are generated around the modulated sampling frequency $A_0\omega_0$ and all its odd harmonics, surrounded by a diminishing series of sidetones separated by a frequency equal to the modulating frequency ω_m, Figure 4.10(a). The profile of this sidetone structure changes considerably as a function of both M and the ratio ω_0/ω_m. In this respect the sidetone structure of PIWM more closely resembles squarewave FM than PWM.

For PIM the minimum permitted ratio of sampling frequency to modulating frequency is 2:1 in order to satisfy sampling theorem requirements. To generate PIWM the PIM pulses have been divided by two using a bistable, reducing the minimum ratio of ω_0/ω_m to unity for PIWM. This does not violate sampling theory requirements however, since a PIWM wavetrain carries information on both rising and falling edges, effectively taking two samples per cycle and permitting economic usage of the spectrum. The interval between adjacent sampling instants is therefore identical for both PIM and PIWM.

In contrast to isochronous techniques such as PWM and PPM the sampling frequency $A_0\omega_0$ is not fixed, but increases as the modulation index M increases. For values of M less than approximately 50%, A_0 may be approximated by $1 + M^2/2$. For example, when $M = 0.2$, $A_0 = 1.02$, shifting the effective sampling frequency upwards by 2%. The magnitude of each sidetone component cannot be calculated by employing a single index m, as is the usual practice, but must be calculated from the combination $n_1 + 2n_2 = m$. Fortunately, the resulting Bessel functions decrease rapidly, making it rarely necessary to use more than three pairs of values for n_1 and n_2 to achieve good accuracy.

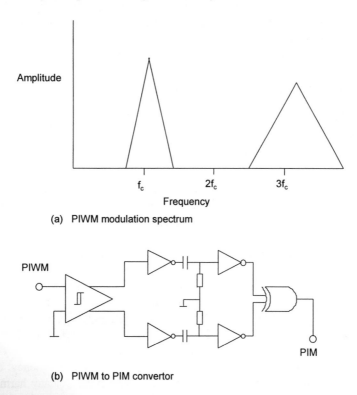

(a) PIWM modulation spectrum

(b) PIWM to PIM convertor

Figure 4.10 *PIWM modulation and conversion*

Demodulation is carried out by first converting the PIWM waveform to PIM pulses and then employing PIM demodulation techniques. This process is facilitated by the use of a complementary output stage within the receiver feeding pairs of logical invertors configured as differentiators followed by an OR gate to recombine the two pulse streams, as in Figure 4.10(b). At low sampling ratios PIWM displays up to 4 dB higher signal-to-noise ratio than PPM under identical conditions, whereas this position is reversed at high sampling ratios [55].

4.3.7 Pulse frequency modulation

In pulse frequency modulation (PFM) the instantaneous frequency of a train of narrow pulses is determined by the modulating signal amplitude. PFM modulation can be conveniently and simply performed by using a voltage controlled multivibrator (VCM) followed by a circuit to generate low duty cycle pulses, Figure 4.11(a). Below carrier frequencies of approximately 20 - 30 MHz standard voltage controlled multivibrators from the TTL logic family are suitable, whereas above this frequency ECL devices become increasingly necessary.

The PFM spectrum for a series of pulses of width τ and repetition frequency ω_c when modulated by a sinewave of frequency ω_m to a frequency deviation $\Delta\omega$ may be represented by [6]:

$$v(t) = \frac{A\omega_c\tau}{2\pi}\left\{1+\frac{2M}{\omega_c\tau}\sin(\omega_m\tau/2)\cos(\omega_m t-\omega_m\tau/2)\right.$$

$$\left. +2\sum_{n=1}^{\infty}\sum_{k=-\infty}^{\infty}J_k(nM)\frac{\sin(n\omega_c+k\omega_m)\tau/2}{n\omega_c\tau/2}\cos\left[(n\omega_c+k\omega_m)t-k\omega_m\tau/2\right]\right\} \qquad (4.6)$$

where $M = (\Delta\omega/\omega_m)$ is the modulation index. As with other frequency modulation techniques, operation with a modulation index greater than unity is possible due to the manner in which the parameter is defined.

Neglecting the DC term, an unmodulated pulse carrier consists of the fundamental and harmonics of the pulse train whose amplitudes follow a $(\sin x)/x$ envelope determined by the pulse width τ. Under modulation conditions the PFM spectrum consists of a baseband component along with a sidetone pattern set around the carrier frequency and all its harmonics, Figure 4.11(b). This sidetone pattern is slightly asymmetrical, with the upper sidetones being stronger then their lower sidetone counterparts. Unlike PWM and PPM the number of sidetones appearing around any particular carrier harmonic is not determined solely by the amplitude of the input signal, but is also a function of input frequency, as with other FM techniques. Since the effective modulation index for any harmonic is proportional to the harmonic number a large degree of sidetone overlap can occur around harmonics at high values of M. PFM can be classified as an anisochronous PTM technique since the carrier frequency component in the modulation spectrum will vanish when the modulation index M is equal to 2.405 (as a consequence of $J_0(2.405) = 0$).

Demodulation is usually achieved by threshold detection and some form of monostable circuit to produce equal length pulses, followed by low-pass filtering to recover the baseband signal component from the PFM modulation spectrum directly [23], Figure 4.11(c). However, a 3 dB improvement in recovered signal-to-noise ratio can be obtained under most conditions by employing double-edge detection and pulse generation [56,57], as illustrated in the next section.

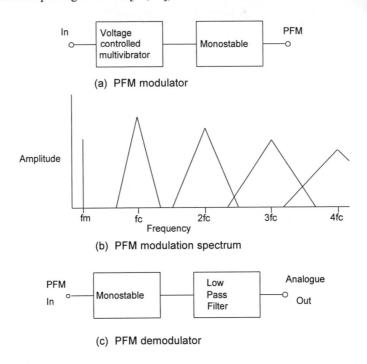

(a) PFM modulator

(b) PFM modulation spectrum

(c) PFM demodulator

Figure 4.11 *Pulse frequency modulation*

In common with other PTM techniques, PFM also displays a noise threshold, below which the noise performance of the system rapidly deteriorates. For PFM operating above threshold, the demodulated noise spectrum is not flat, but contains an element inversely proportional to the square of the input signal frequency [58,59], as with other frequency modulation techniques.

4.3.8 Squarewave frequency modulation

Squarewave frequency modulation is an anisochronous PTM technique closely related to PFM and is the pulse equivalent of sinewave frequency modulation (FM); essentially consisting, as previously shown in Figure 4.3(d), of a series of squarewave edge transitions occurring at the zero crossing points of FM [60]. The modulation spectrum of SWFM is composed of an FM-like spectrum at the carrier fundamental frequency ω_c with slightly modified forms at all odd harmonics, Figure 4.12(a). The sideband pattern around the *n*th harmonic of the carrier frequency displays a frequency deviation of *n* times the deviation around the carrier fundamental. Thus the spectral spreading around each carrier harmonic can be calculated from Carson's rule if desired. The modulation spectrum with an input frequency ω_m may be expressed analytically as [25,28]:

$$s(t) = AD \sum_{n=-\infty}^{\infty} \text{sinc}(n\pi D) \sum_{k=-\infty}^{\infty} J_k(nM)\exp\left[j(n\omega_c + k\omega_m)t\right] \qquad (4.7)$$

where A is the pulse amplitude, D is the duty cycle, M is the modulation index and J_k is a Bessel function of the first kind, order k.

When the carrier wave is a perfect squarewave with 50% duty cycle *(D = 0.5)* there is no baseband component to the spectrum. However, departure from an accurate 50% duty cycle will result in the generation of a baseband component as well as noticeable FM sidebands at even harmonics of ω_c along with an increasingly skewed sideband amplitude distribution. Extreme asymmetry in the carrier wave duty cycle produces a modulation spectrum very similar to that of PFM where the relative power of the higher order terms increases with decreasing pulse width, indicating that PFM requires a wider transmission bandwidth than SWFM to achieve the same signal-to-noise ratio. By virtue of the carrier null experienced when the modulation index M is equal to 2.405, there exists a well defined point for convenient SWFM system calibration.

SWFM modulation can be conveniently and easily performed by using a voltage-controlled multivibrator (VCM) with a squarewave output, preceded by a low-pass filter to bandlimit the modulating signal.

Methods for demodulation may be deduced directly from the SWFM modulation spectrum. Ordinary FM demodulation techniques may be adopted, locked on to the carrier or harmonics using a phase-locked loop (PLL). However, limited linearity and noise performance usually result from this approach as a consequence of selecting only a restricted spectral slice. Of more interest is the approach based on pulse regeneration, where either the leading or trailing edges, or preferably both, of the SWFM waveform are used to reconstruct the associated PFM modulation spectrum, since this contains a baseband component that can then be selected directly by a low-pass filter.

PFM spectral reconstruction from single-edge detection produces a spectrum similar to Figure 4.11(b) in which, along with the desired baseband component, there are sideband structures around every harmonic of the carrier frequency. In contrast, for double-edge detection, only the even harmonics are present above the baseband region, Figure 4.12(b). This produces greater spectral efficiency via a lower bandwidth overhead penalty by lowering the minimum SWFM carrier frequency ω_c necessary for any particular maximum modulating signal frequency ω_m. In addition, there is a doubling of the reconstructed PFM modulation index resulting in superior signal-to-noise performance compared to the single-edge case [56].

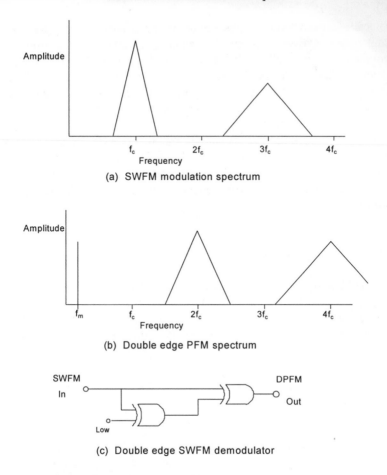

(a) SWFM modulation spectrum

(b) Double edge PFM spectrum

(c) Double edge SWFM demodulator

Figure 4.12 *Squarewave frequency modulation*

Practical circuit implementation of double-edge baseband pulse regeneration has been found to be particularly simple by employing a number of exclusive OR logic gates in a logic-rectification and delay circuit [25], rather than the usual monostable approach, as in Figure 4.12(c).   Again, TTL logic families are adequate for carrier frequencies up to approximately 20 - 30 MHz, but beyond this value ECL devices must be employed.

4.3.8.1 Extended squarewave frequency modulation
It is also possible to employ SWFM in fibre systems intended for high-precision applications, such as transmission of DC-coupled wideband instrumentation signals where extremely low DC drift is often essential, usually with fibre lengths of around 1km or less (Figure 4.13). Frequency drift of the voltage-controlled oscillator (multivibrator) has previously set the resolution limit for SWFM, even

after localised temperature compensation schemes have been applied to both transmitter and receiver. However, by adopting an approach based on classical control theory to servo out the drift component by the use of a separate feedback- or feedforward-derived signal and integrator loop, dramatically improved resolution and temperature stability can be obtained along with improved low-frequency linearity and unimpaired high-frequency performance [60,61]. The additional low-frequency signal path may be provided by a second fibre in the case of the feedback-derived signal, or by multiplexing in the case of the feedforward-derived system. An additional benefit is that all transmitters become interchangable with all receivers since servo action automatically minimises DC offset in both units to provide premium performance without individual calibration.

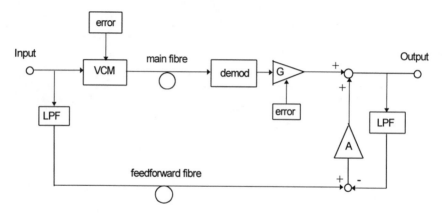

Figure 4.13 *SWFM with extended response*

4.3.9 Hybrid pulse time modulation

Very little work has been reported on transmission of PTM multiplexes. By virtue of their isochronous nature, both PWM and PPM can be transmitted using time division multiplexing (TDM), whereas frequency division multiplexing (FDM) is more suited to anisochronous PTM techniques. Dedicated two-channel PTM systems have, however, received limited attention [62], with a particularly simple two-channel approach being formed by cascading a PFM and a PWM modulator, as in Figure 4.14, to produce a hybrid or compound PTM pulse train, such that the normally constant width of a PFM pulse is modulated by a second signal [63-66]. Demodulation is performed by edge separation to reconstruct the original PFM and PWM pulse streams before recovery of the two original signals, but not without a certain degree of crosstalk [64].

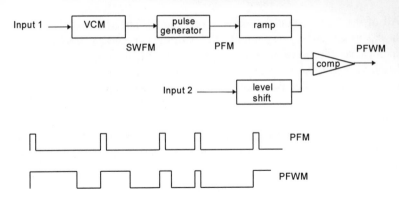

Figure 4.14   *Hybrid pulse frequency and width modulation*

4.4 PTM performance potential

The high-speed performance potential of PTM techniques may be explored by reference to the interactions between spectral occupancy, distortion and signal-to-noise ratio. Suitability of a particular technique for transmission of analogue data can be evaluated by its ability to produce low-distortion concurrent with a low sampling ratio and a high signal-to-noise ratio. From the standpoint of spectral overlap between the modulating signal and the carrier fundamental lower sidetone structure we can predict the minimum carrier frequency for each of the PTM methods [67].

Since all PTM methods operate via time-amplitude conversion, jitter in the transmission system is of concern to the system designer. In general, pulse jitter generated in a PTM modulator and optical transmitter is negligible compared to contributions from other sources [24]. The main jitter contribution is through shot noise and thermal noise in the optical detector and receiver lowering the carrier-to-noise ratio at the demodulator. Above the threshold where edge noise begins to dominate impulse noise, that is beyond a carrier-to-noise ratio of around 20dB [58], PTM pulse slicing and regeneration should occur at half pulse amplitude for minimum jitter penalty through symmetrical eye closure. This may not be the case for an active network that includes fibre amplifiers or optically amplified receivers, where a lower decision threshold may be appropriate due to the asymmetrical eye pattern used to counter effects of signal-dependent noise through amplifier spontaneous emission [68]. As with most other high-speed fibre systems, transimpedance receiver topologies tend to dominate due to the higher bandwidths and absence of equalisation problems, even though there is a small noise penalty from the feedback resistor. The gain of an APD photodetector in a high-sensitivity system must be balanced as usual against its excess noise to minimise the overall noise input to the demodulator.

All of the PTM techniques share a very similar spectral modulation structure with sidetones appearing around the carrier (sampling) frequency and its harmonics, as discussed in a previous section. However, there are particular differences between the techniques concerning the presence of a baseband component and detailed behaviour of the carrier frequency lower sidetone profile which will determine the minimum sampling ratio for a specified distortion behaviour.

PWM and narrowband PPM share an identical sidetone structure around the sampling fundamental, resulting in a ratio of recovered signal modulation level A_m to sidetone amplitude A_k of:

$$A_m/A_k = \pi M/2J_k(\pi M) \qquad (4.8)$$

where $J_k(\beta)$ is a Bessel function of the first kind, order k and argument β. M is the modulation index.

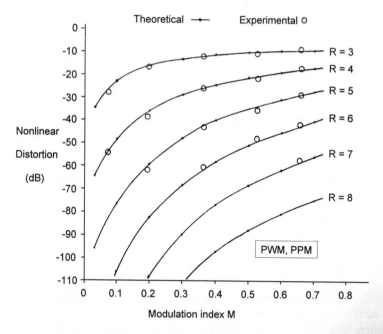

Figure 4.15 *Distortion behaviour of PWM and PPM*

From this A_m/A_k can be plotted as a function of M for a range of sampling ratios R, given by f_c/f_m, to gain insight into the limit on distortion imposed by intrusion of the sidetone structure into the recovered baseband. Figure 4.15 illustrates that to obtain a distortion level of 1% (-40 dB), for example, in the recovered signal at its maximum frequency, a minimum sampling ratio R of 4 at

10% modulation index is required, rising to 6 at an index of 70%. (A modulation index of 70% - 80% is typically the maximum value feasible in PTM modulators before the onset of additional circuit non-linearities). This data may be recast as in Figure 4.16 to produce design curves indicating the minimum sampling ratio required for specific levels of distortion. Clearly, both PWM and PPM are unlikely to be operated with sampling ratios below 6 - 7 to obtain low distortion behaviour. Their main attribute is that of circuit simplicity.

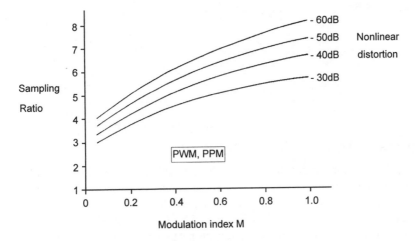

Figure 4.16 *Sampling ratios for isochronous PTM*

For PFM the ratio of recovered signal component to carrier sidetone level may be expressed as:

$$A_m/A_k = \pi \Delta f \tau / \sin[\pi(f_c + kf_m)\tau] J_k(M) \tag{4.9}$$

where Δf is the carrier frequency excursion (also expressible as Mf_m) and τ is the PFM pulse width, with $k=R-1$.

Figure 4.17 illustrates the behaviour of PFM in this respect, showing that at around a modulation index of 0.5 a minimum sampling ratio of 5 is necessary to obtain 1% distortion level, increasing to 7 at $M = 2.0$. From the standpoint of bandwidth utilisation for a specified level of nonlinear distortion, PFM modulation is marginally superior to PWM and PPM.

For SWFM with double-edge pulse regeneration demodulation we may write the ratio of recovered signal to sidetone level as [25]:

$$A_m/A_k = 2\pi \Delta f \tau / \sin[\pi(2f_c + kf_m)\tau] J_k(2M) \tag{4.10}$$

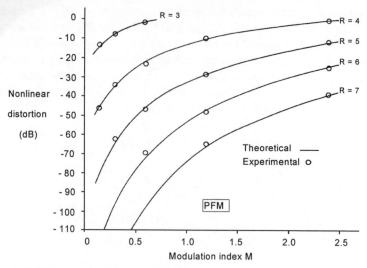

Figure 4.17 *Distortion behaviour of PFM*

Equation 4.10 for SWFM differs from PFM Equation 4.9 in that the sidetones are now generated around the carrier second harmonic $2f_c$ with complete suppression of the spectral structure around the carrier fundamental for a unity mark/space ratio carrier wave. In addition the recovered baseband signal level is twice that of PFM. This results in a superior distortion performance from SWFM for a given sampling ratio, as demonstrated in Figure 4.18, producing a minimum sampling ratio of 3 for 1% distortion at $M = 0.5$, rising to 5 at $M = 2.0$.

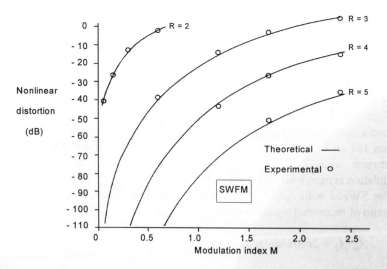

Figure 4.18 *Distortion behaviour of SWFM*

A comparison between SWFM and PFM is drawn in Figure 4.19, where the minimum required sampling ratio is plotted for a range of distortion performance levels as a function of modulation index. SWFM is clearly markedly superior to PFM, in this respect, achieving theoretical distortion levels due to sidetone incursion of better than -60 dB under modulation conditions where PFM can only achieve -30 dB. In turn PFM is superior to PWM under similar modulation conditions, making SWFM the best choice from the standpoint of minimum sampling ratio and bandwidth overhead.

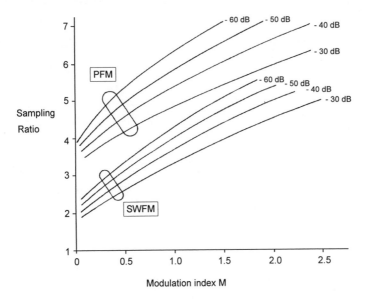

Figure 4.19 *Sampling ratios for anisochronous PTM*

In addition to the trade-off between the minimum carrier frequency and distortion performance we are also interested in the interaction between the signal-to-noise performance and the received pulse edge speed, as dictated by the transmission channel bandwidth B_t. The unweighted improvement factor *IF*, given by signal-to-noise ratio SNR minus carrier-to-noise ratio CNR, expressed in dBs, varies above threshold as the square of the ratio of transmission channel bandwidth to carrier frequency. The constant of proportionality is slightly different for each of the PTM techniques [25], such that:

PWM $\qquad IF = \pi^2/8(mB_t/f_c)^2$ $\qquad\qquad$ (4.11)

PPM $\qquad IF = \pi^2/4(mB_t/f_c)^2$ $\qquad\qquad$ (4.12)

PFM $\qquad IF = 3/4(mB_t/f_c)^2$ $\qquad\qquad$ (4.13)

SWFM $IF = 3/2(mB_t/f_c)^2$ (4.14)

To provide an illustrative comparison, Figure 4.20 highlights the improvement factor achievable from each of the PTM techniques under the constraint of limiting the transmission channel bandwidth B_t to be 1 GHz, a factor of 10 higher than the maximum signal frequency f_m of 100 MHz in this example, along with a PPM and PFM pulse width of 100 ps.

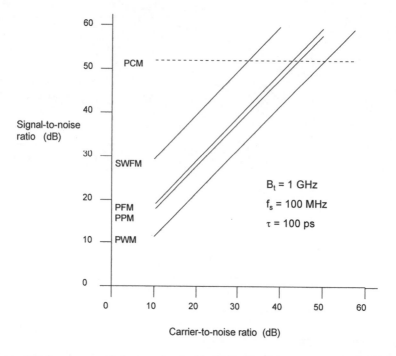

Figure 4.20 *Improvement factors for the PTM family*

This method of presentation is informative since it takes into account directly the different sampling ratio behaviour of each technique. A distortion level of -40 dB is taken as the criterion for choosing the appropriate modulation index and sampling ratio for each technique. The theoretical performance of 8-bit PCM above threshold is also included for comparison, demonstrating that at higher carrier-to-noise ratios PTM methods become progressively superior.

At a modulation index of 0.5 with the minimum permitted sampling ratio of 5, PWM can only achieve an improvement factor of just under 2 dB. Under the same conditions PPM is able to achieve 7.8 dB with PFM returning a marginally better result of 9.5 dB with a modulation index of 1.1 and a sampling ratio of 5.5. Owing to the lower sampling ratio required by SWFM for the same level of distortion performance this technique is able to achieve a superior improvement factor of 20 dB with a modulation index of 0.9 and a sampling ratio of 3.5, making it the preferred PTM choice when transmission bandwidth is at a premium.

4.5 Summary

This chapter has surveyed and reviewed the range of PTM techniques available for fibre transmission and outlined their advantages from the standpoint of bandwidth-efficient modulation of analogue-sourced signals. A general classification method has been described by which different members of the PTM family may be categorised. Equations have been presented to describe the modulation spectrum for each of the different methods. To explore the high-frequency potential of PTM the interaction between distortion performance and minimum sampling ratio as a result of sidetone overlap has been examined. Calculations have indicated that, operating above threshold, SWFM could, for example, offer an improvement factor of almost 20 dB for a system restricted to a transmission channel bandwidth of 1 GHz operating with a maximum input signal frequency of 100 MHz.

4.6 References

1   Olshansky, R., and Lanzisera, V.A.: '60-channel FM video subcarrier multiplexed optical communication system', *Electron. Lett.*, 1987, **23**, pp. 1196-1198

2   Salgado, H.M., and O'Reilly, J.J.: 'Performance assessment of FM broadcast subcarrier multiplexed optical systems', *IEE Proc.-J*, 1992, **140**, pp. 397-403

3   Wilson, B., and Ghassemlooy, Z.: 'Pulse time modulation techniques for optical communications: a review", *IEE Proc.-J*, 1993, **140**, pp. 346-357

4   Brace, D.J., and Heatley, D.J.: 'The application of pulse modulation schemes for wideband distribution to customers' premises', *Proc. 6th European Conference on Optical Communications*, York, 1980, pp. 446-449

5   Cooke, D., Jelonek, Z., Oxford, A.J. and Fitch, E.: 'Pulse Communication', *J. IEE*, 1947, **94**, Part IIIA, pp. 83-105

6   Fitch, E.: 'The spectrum of modulated pulses', *J. IEE*, 1947, **94**, Part IIIA, pp. 556-564

7   Jelonek, Z.: 'Noise problems in pulse communication', *J. IEE*, 1947, **94**, Part IIIA, pp. 533-545

8   Levy, M.M.: 'Some theoretical and practical considerations of pulse modulation', *J. IEE*, 1947, **94**, Part IIIA, pp. 565-572

9   Parks, G.H., and Moss, S.H.: 'A new method of wide-band modulation of pulses', *J. IEE*, 1947, **94**, Part III A, pp. 511-516

10 Moss, S.H., and Parks, G.H.: 'Pulse communication on lines', *J. IEE*, 1947, **94** Part IIIA, pp. 503-510

11 Schrocks, C.B.: 'Proposal for a hub controlled cable television system using optical fiber', *IEEE Trans. on Cable Television*, 1979, **CATV-4**, pp. 70-77

12 Berry, M.C., and Arnold, J.M.: 'Pulse width modulation for optical fibre transmission of video', IEE Int. Conf. on the Impact of VLSI Technology on Communication Systems, London, 1983

13 Suh, S.Y.: 'Pulse width modulation for analog fiber-optic communications', *IEEE J. Lightwave Technol.,* 1987, **LT-5**, pp. 102-112

14 Wilson, B., and Ghassemlooy, Z.: 'Optical pulse width modulation for electrically isolated analogue transmission', *J. Phys. (E)*, 1985, **18**, pp. 954-958

15 Wilson, B., and Ghassemlooy, Z.: 'Optical PWM data link for high quality analogue and video signals', *J. Phys. (E)*, 1987, **20**, pp. 841-845

16 Wilson, B., and Ghassemlooy, Z.: 'Optical fibre transmission of multiplexed video signals using PWM', *Int. J. of Optoelectronics*, 1989, **4**, pp. 3-17

17 Nicholson, G.: 'Modulation techniques for cable television distribution on optical fibres', *Australia Telecom Res.*, 1983, **17**, pp. 25-37

18 Matsubara, Y.I., Li, J., Yoshida, H., Shinohara, S., Tsuchiya, E., and Ikeda, H.: 'Pulse-width modulated video signal transmission through optical fibre, *Proc. 35th Midwest Symposium on CCTs & Systems*, Washington USA, 9th-12th August 1992, **1**, pp. 357-360

19 Heatley, D.J.T.: 'Video transmission in optical fibre local networks using pulse time modulation', ECOC 83 - 9th European Conference on Optical Comms, Geneva, September 1983, pp. 343-346

20 Okazaki, A.: 'Still picture transmission by pulse interval modulation', *IEEE Trans. on Cable Television*, 1979, **CATV-4,** pp. 17-22

21 Heatley, D.J.T.: 'Unrepeatered video transmission using pulse frequency modulation over 100 km of monomode optical fibre', *Electron. Lett.*, 1982, **18**, pp. 369-371

22 Heatley, D.J.T., and Hodgkinson, T.G.: 'Video transmission over cabled monomode fibre at 1.5 mm using PFM with 2-PSK heterodyne detection', *Electron. Lett.*, 1984, **20**, pp. 110-112

23 Heker, S.F., Herskowitz, G.J., Grebel, H. and Wichansky, H.: 'Video transmission in optical fiber communication systems using pulse frequency modulation', *IEEE Trans. on Comms.*, 1988, **36**, pp. 191-194

24 Kanada, T., Hakoda, K., and Yoneda, E.: 'SNR fluctuation and non-linear distortion in PFM optical NTSC video transmission systems', *IEEE Trans. on Comms.*, 1982, **COM-30**, pp. 1868-1875

25 Lu, C.: 'Optical transmission of wideband video signals using SWFM', Ph.D. Thesis, University of Manchester Institute of Science and Technology, Manchester UK, 1990

26 Pophillat, L.: 'Video Transmission using a 1.3 μm LED and Monomode Fiber', 10th European Conf. on Optical Comms., Stuttgart, W. Germany, 1984, pp. 238-239

27   Sato, K., Aoygai, S., and Kitami, T.: 'Fiber optic video transmission employing square wave frequency modulation', *IEEE Trans. on Comms.*, 1985, **COM-33**, pp. 417-423

28   Wilson, B., Ghassemlooy, Z., Darwazeh, I., Lu, C., and Chan, D.: 'Optical squarewave frequency modulation for wideband instrumentation and video signals', IEE Colloquium on Analogue Optical Comms., London, 1989, Digest 1989/165, Paper 9

29   Wilson, B., Ghassemlooy, Z., and Lu, C.: 'Optical fibre transmission of high-definition television signals using squarewave frequency modulation', Third Bangor Symp. on Comms., University of Wales - Bangor, May 1991, pp. 258-262

30   Wilson, B., Ghassemlooy, Z., and Lu, C.: 'Squarewave FM optical fibre transmission for high-definition television signals', (Fibre Optics 90, 1990, London), *Proc. Int. Soc. Optical Eng.*, 1990, **1314**, pp. 90-97

31   Okazaki, A.: 'Pulse interval modulation applicable to narrowband transmission', *IEEE Trans. on Cable Television*, 1978, **CATV-3**, pp. 155-164

32   Sato, M., Murata, M., and Namekawa, T.: 'Pulse interval and width modulation for video transmission, *IEEE Trans. on Cable Television*, 1978, **CATV-3**, pp. 166-173

33   Sato, M., Murata, M., and Namekawa, T.: 'A new optical communication system using the pulse interval and width modulated code', *IEEE Trans. on Cable Television*, 1979, **CATV-4**, pp. 1-9

34   Wilson, B., Ghassemlooy, Z., and Cheung, J.C.S.: 'Spectral predictions for pulse interval and width modulation', *Electron. Lett.*, 1991, **27**, pp. 580-581

35   Calvert, N.M., Sibley, M.J.N., and Unwin, R.T.: 'Experimental optical fibre digital pulse position modulation system', *Electron. Lett.*, 1988, **24**, pp. 129-131

36   Cryan, R.A., Unwin, R.T., Garratt, I., Sibley, M.J.N., and Calvert, N.M.: 'Optical fibre digital pulse-position-modulation assuming a Gaussian received pulse shape', *IEE Proc.-J*, 1990, **137**, pp. 89-96.

37   Garrett, I.: 'Pulse position modulation for transmission over optical fibres with direct or hetrodyne detection', *IEEE Trans. on Comms.*, 1983, **COM-31**, pp. 518-526

38   Black, H.S.: 'Modulation Theory', chapter 17 (Van Nostrand, New York, 1953)

39   Bennet, W.R.: 'New results in the calculation of modulation components', *Bell Syst. Tech. J.*, 1933, **12**, pp. 228-243

40   Stuart, R.D.: 'An Introduction to Fourier Analysis', chapter 6 (Chapman and Hall, London, 1962)

41   Barton, T.H.: 'Pulse width modulation waveforms; the Bessel approximation', *IEEE Industry Application Soc.*, 1st-5th October 1978, Ontario, pp. 1125-1130

42   Marvasti, F., Analoui, F.M., and Gamshadzahi, M.: 'Recovery of signals from nonuniform samples using iterative methods' *IEEE Trans. on Signal Processing*, 1991, **39**, pp. 872-878

43   Wilson, B. and Ghassemlooy, Z.: 'Multiple sidetone structure in pulse width modulation', *Electron. Lett.*, 1988, **24**, pp. 516-518

44   Wilson, B., Ghassemlooy, Z., and Lok, A.: 'Spectral structure of multitone pulse width modulation', *Electron. Lett.*, 1991, **27**, pp. 702-704

45   Wilson, B. and Ghassemlooy, Z.: 'Differential phase error in pulse width modulated TV transmission systems', *Electron. Lett.*, 1987, **23**, pp. 1133-1134

46   Biase, V.D., Passeri, P., and Pietroiusti, R.: 'Pulse analog transmission of TV signal on optical fibre', *Alta Frequenza*, 1987, **LVI-N4**, pp. 195-203

47   Holden, W.S.: 'An optical-frequency pulse-position-modulation experiment', *Bell Syst. Tech. J.*, 1975, **54**, pp. 285-296

48   Uneo, Y., and Yasugi, T.: 'Optical fiber communication system using pulse-interval modulation', *NEC Research & Development J.*, 1978, **48**, pp. 45-52

49   Fyath, R.S., Abdullah, S.A., and Glass, A.M.: 'Spectrum investigation of pulse interval modulation', *Int. J. Electron.*, 1985, **59**, pp. 597-601

50   Tripathi, J.N.: Spectrum measurement of pulse-interval modulation', *Int. J. Electron.*, 1980, **49**, pp. 415-419

51   Sharma, P.D., and Tripathi, J.N.: 'Signal to noise ratio studies of PIM and PIM-FM systems', *Int. J. Electron.*, 1970, **28**, pp. 129-141

52   Schwartz, M., Bennett, W.R., and Stein, S.: 'Communication Systems and Techniques', pp. 156-158 (McGraw-Hill, NewYork, 1966)

53   Marougi, S.D., and Sayhood, K.H.: Noise performance of pulse interval modulation systems', *Int. J. Electron.*, 1983, **55**, pp. 603-614

54   Wilson, B., Ghassemlooy, Z., and Cheung, J.C.S.: 'Optical pulse interval and width modulation for analogue fibre communications', *IEE Proc.-J*, 1992, **139**, pp. 376-382

55   Marougi, S.D., and Sayhood, K.H.: 'Signal-to-noise performance of the pulse interval and width modulation', *Electron. Lett.*, 1983, **19**, pp. 528-530

56   Heatley, D.J.T.: 'SNR comparison between two designs of PFM demodulator used to demodulate PFM or FM', *Electron. Lett.*, 1985, **21**, pp. 214-215

57   Nakamura, M., Kaito, K., and Ozeki, T.: 'Modal noise reduced PFM transmission by monopulse to twin-pulse conversion', *Electron. Lett.*, 1985, **21**, pp. 307-308

58   Drukarve, A.I.: 'Noise performance and SNR threshold in PFM', *IEEE Trans. on Comms.*, 1985, **COM-33**, pp. 708-711

59   Webb, R.P.: 'Output noise spectrum from demodulator in an optical PFM system', *Electron. Lett.*, July 1982, **18**, pp. 634-636

60   Wilson, B., Ghassemlooy, Z., and Lu, C.: 'Squarewave frequency modulation techniques', *IEEE Trans. on Comms.*, Accepted for publication

61   Wilson, B., and Ghassemlooy, Z.: 'DC to wideband modulation in fibre systems', *Proc. Int. Soc. Optical Eng*, 1993, **1974**, pp. 90-98

62   Sato, M., Murata, T., and Namekawa, T.: 'A simplified voice, two-channel transmission experiment by pulse interval-width modulation', *IEICE Trans. Comms.*, 1979, **J62-B**, pp. 712-713

63   Tanaka, T., and Okamura, N.: 'Multiple transmission systems using PFM and PWM', *Electron. and Comms. in Japan*, 1991, Part 1, **74**, pp. 65-80

64   Ghassemlooy, Z., Wilson, B., and Lok, A..: 'Hybrid pulse time modulation techniques', *Proc. Int. Soc. Opt. Eng.*, 1993, **1974**, pp. 120-125

65   Ghassemlooy, Z., and Wilson, B.: 'Optical compound pulse time modulation for analogue fibre transmission', *Proc. IEEE SICON/ICIE '93*, September 1993, Singapore, **2**, pp. 630 - 634

66 Ghassemlooy, Z., Issa, A.A, and Wilson, B.: 'Spectral predictions for hybrid pulse frequency and width modulation', *Electron. Lett.*, 1994, **30**, pp. 933-935

67 Wilson, B., Ghassemlooy, Z., and Lu, C.: 'High-speed pulse time modulation techniques', *Proc. Int. Soc. Optical Eng.*, 1992, **1787**, pp. 292-302

68 Lane, P.M., Watkins, L.R., and O'Reilly, J.J.: 'Distributed microwave filter realisation providing close to optimum performance for multi-gigabit optical communications', *IEE Proc.-J*, 1992, **139**, pp. 280-287

Chapter 5

Soliton pulse position modulation

J. M. Arnold

5.1 Introduction

The use of the nonlinear phenomenon of optical fibre solitons for the transmission of digital information has recently received a great deal of attention [1, 2, 3], made possible by the recent development of optical amplifiers which counteract the fibre loss. A soliton is an optical pulse in which the nonlinearity of the fibre exactly compensates its natural dispersion, and an isolated soliton will propagate without dispersion over indefinitely large distances. It would therefore seem that solitons are attractive vehicles for the transmission of data. In principle, the capacity of such a communication system should be extremely large; optical fibre solitons of pulse widths on the order of 100 fs have been reported, which, if conventional communications theory were observed in this case, would suggest attainable bit-rates of several thousand Gbs^{-1} over arbitrarily large transmission distances. However, a number of practical limitations enter the assessment of this concept. Firstly, the maximum transmission rate is not as simply related to the fundamental pulse width as would be the case in a linear communication system; a nonlinear interaction exists between successive pulses in a train of solitons with the effect that the pulses must be separated by at least several pulse widths in order to mitigate the effects of the interaction, and the maintenance of the interaction at sufficiently low levels requires a limitation on the transmission distance [4]. Secondly, the production of a soliton requires a relatively large amount of energy per pulse. Roughly speaking, a 10 ps (FWHM) soliton requires on the order of 0.26 pJ of energy in a low-dispersion fibre of dispersion parameter $D = -3.7$ ps^2 km^{-1} and effective core area 25 μm^2, and the pulse energy scales inversely with the duration of the soliton (see below); thus a 50 ps period train of 5 ps pulses requires approximately 10 mW of average optical power. Because of this relationship between energy-per-bit and pulse width, a soliton communication system trades signal power for bandwidth in a very direct way. Thirdly, the effects of noise in nonlinear communications channels are considerably more complicated than in the linear case. In particular, the transmission of solitons over a very long fibre path requires periodic amplification to compensate for the dissipation of the solitons by the fibre attenuation; the amplification process introduces noise which interacts with the solitons and affects the nonlinear propagation process.

The phenomenon of a train of optical fibre solitons, modulated in some way by impressed disturbances, is increasingly being identified as a new type of dynamical system, with interesting dynamical behaviour which can be studied both experimentally and mathematically. This perception leads rather naturally to an investigation of the behaviour of this system when subjected to various forms of pulse-analogue modulation, to which this chapter is devoted. The principal motivation for the study of solitons in pulse position modulation (PPM) arises from the fact that PPM with short pulses permits a greater transmission rate than longer pulses, and intense pulses exhibit higher signal-to-noise ratio than weak pulses. Consequently, intense short pulses are desirable for PPM, but fibre dispersion limits the minimum width an individual pulse can have after linear transmission. This consideration leads naturally to speculation on solitons in PPM because the soliton does not suffer dispersion. However, this is only one example of a general interest which is developing in the understanding of dynamical phenomena involving solitons, to which this work is addressed.

In the following sections the theory of pulse position modulation and noise will be developed for the soliton channel, using a linearised lattice approximation to the underlying nonlinear equations which describe the soliton propagation. The analysis is not restricted to linear phenomena, however, and nonlinear behaviour can also be exhibited by the lattice approximation, as is shown in section 5.5. The basic approach is exemplified by a treatment of pulse position modulation in the small-deviation case; this permits the linearisation of the dynamical equations, which are similar to those of the Toda lattice. Noise is also a linear perturbation, and can be very effectively treated in this approximation. The dominant noise source in very long soliton communication systems is amplified spontaneous emission (ASE) produced by the optical amplifiers; also, mutual interpulse interference is caused by the nonlinear interaction which exists between solitons in a multiple pulse train [4]. The intersoliton interaction turns the soliton pulse train into a dynamical system in its own right, and this dynamical system may be stable or unstable, depending on whether the pulse interaction is repulsive or attractive, respectively. In the stable case, modulation impressed on one soliton will disperse onto neighbouring solitons, causing a noise component to develop as the pulse train propagates. An expression is given here for the collective effects of that interference. In an interacting pulse train the ASE noise evolves on long range scales differently to its behaviour in a single isolated soliton pulse. A further expression is obtained for the evolution of ASE in the presence of intersoliton interactions in an infinite pulse train, which generalises a well-known result of Gordon and Haus [10] for the case of an isolated soliton, and which exhibits significant differences from that formula.

5.2 Nonlinear pulse propagation in optical fibres

The propagation of an optical fibre soliton along a nonlinear lossless optical fibre obeys the nonlinear Schrodinger equation (NLSE)

$$i\partial_x\phi + \tfrac{1}{2}\partial_t^2\phi + |\phi|^2\phi = 0 \tag{5.1}$$

where ϕ is the complex envelope of the propagating mode, and normalised 'soliton' units are used for the dimensions of x (distance along the fibre) and t (local time in a frame moving along the fibre at the modal group velocity) [5]. (The soliton units are defined as follows: if 1 soliton time unit is t seconds, and 1 soliton distance unit is L_0 metres, the two quantities are related by $\tau^2/L_0 = |D|$, where D is the dispersion parameter of the fibre. Thus, for $\tau = 5.7$ ps, corresponding to a soliton FWHM pulse width of 10 ps, and $D = -3.7$ ps^2 km^{-1}, then $L_0 = 8.7$ km. Physical units can be recovered from normalised units by scaling distances by L_0 and times by τ). It has been established that for a fibre with weak loss which is periodically exactly compensated by ideal linear amplifiers Equation 5.1 can be interpreted as applying to certain average quantities, rather than to the nonlinear wave itself [6]. In practice the amplification is provided by erbium-doped fibre amplifiers (EDFA).

The 1-soliton solution of Equation 5.1 is

$$\phi = e^{ix/2}\sec ht \tag{5.2}$$

having a full-width at half-maximum intensity (FWHM) of $\tau_s = 1.76$ soliton units, or 1.76τ seconds. This pulse propagates without change of shape in a lossless fibre. Other solutions are found by rescaling $\phi\rightarrow\eta\phi$, $t\rightarrow ht$, $x\rightarrow\eta^2 x$, with some arbitrary real parameter η. In physical units the energy of a soliton is $E = 2P\tau$ (Joules), where P is the peak power (Watts) of the optical pulse. A silica fibre with $D = -3.7$ ps^2 km^{-1} requires $P = 23$ mW for a soliton unit $\tau = 5.7$ ps (FWHM $\tau_s = 10$ ps), and hence a pulse energy $E = 0.26$ pJ. Energies for other pulse widths can be obtained by scaling; energy scales inversely with pulse width. Expressions in which time and length are normalised can be converted to physical units by the prescription $t\rightarrow t/\tau$ and $x\rightarrow x/L_0$.

A sequence of solitons in time, launched on the fibre at $x = 0$, can be represented by

$$\phi|_{x=0} = \sum_{k\in Z}\eta_k e^{i\alpha_k}e^{-i\omega_k(t-t_k)}\sec h[\eta_k(t-t_k)] \tag{5.3}$$

where Z represents the set of integers and η_k, α_k, ω_k and t_k are the amplitude, phase, frequency and time respectively of the kth soliton. Any of these four parameters may be used to carry a modulation encoding the sample values of an analogue signal; however, it turns out that these variables are linked in pairs by the underlying dynamics of the soliton field. At the lowest order of approximation, conjugate pairs; are (α_k, η_k) and (t_k, p_k), where the 'momentum' of the pulse is defined by $p_k = -\omega_k \eta_k$.

For pulse analogue communication purposes the case of amplitude modulation (of η_k) can be excluded from the start because according to Equation 5.3 this would require both the amplitude and the width of the solitons to be simultaneously modulated in order to maintain the pulses as solitons, which seems impractical with currently available optical technology at picosecond time scales. (However, digital on-off modulation (ASK) remains practical, since solitons corresponding to zeroes simply do not appear in the sum Equation 5.3). Frequency modulation (of ω_k) alters the velocity of each pulse, which would be undesirable for a long-distance communication system because it induces relative motion of the solitons; it is this relative motion, when established by noise rather than by a signal, which proves to be the dominant noise source in fibre soliton communications. Consequently, we concentrate here on analogue PPM (t_k) as an information-bearing modulation. Pulse phase modulation (PPhM) is also possible, and the analysis follows similar lines to that outlined here for PPM. However, noise affects all degrees of freedom of the soliton pulse train.

It can be shown that the initial conditions in Equation 5.3 at $x = 0$ will evolve along x, to a good approximation, simply by changing the parameters α_k, η_k, p_k and t_k, which become functions of x. Since these parameters carry the modulation, this implies that the propagation of the solitons alters the signal which is encoded into these parameters, resulting in distorted transmission; it is of fundamental importance to characterise this distortion. It has been shown [4] that for two isolated solitons, launched with equal velocities $p_k|_{x=0} = 0$ and unity amplitudes $\eta_k|_{x=0} = 1$, the parameters evolve approximately according to two coupled nonlinear ordinary differential equations ($k \in \{1,2\}$)

$$d_x^2 t_k = -4(-1)^k e^{-(t_2 - t_1)} \cos(\alpha_2 - \alpha_1) \qquad (5.4a)$$
$$d_x^2 \alpha_k = 4(-1)^k e^{-(t_2 - t_1)} \sin(\alpha_2 - \alpha_1) \qquad (5.4b)$$

which have to be solved with initial conditions at $x = 0$ consisting of prescribed values for $\{\alpha_k\}$, $\{t_k\}$, $\{d_x \alpha_k\}$ and $\{d_x t_k\}$. There are also two auxiliary equations which link the conjugate variables

$$d_x \alpha_k = \eta_k^2 / 2 \qquad (5.4c)$$
$$d_x t_k = p_k \qquad (5.4d)$$

Defining

$$q = t_2 - t_1 \qquad (5.5a)$$

$$\alpha = \alpha_2 - \alpha_1 \tag{5.5b}$$

Equations 5.4 a,b become

$$d_x^2 q = -8e^{-q} \cos(\alpha) \tag{5.6a}$$
$$d_x^2 \alpha = 8e^{-q} \sin(\alpha) \tag{5.6b}$$

The following facts can be deduced from Equations 5.6 a,b and initial conditions $d_x q|_{x=0} = 0$ and $d_x a|_{x=0} = 0$ without directly solving them:

(i)   if $\alpha|_{x=0} = 0$ or π and $d_x a|_{x=0} = 0$ then α is a constant of the motion;
(ii)   if $\alpha = 0$ then q oscillates about $q = 0$ with increasing x;
(iii)   if $\alpha = \pi$ then q increases monotonically with increasing x.

The properties (ii) and (iii) are sometimes interpreted as the existence of an attractive ($\alpha = 0$) or repulsive ($\alpha = \pi$) 'force' between the solitons, on account of the similarity between Equation 5.6 and Newton's law for a particle moving in a potential. It was further shown by Gordon [4] that in case (ii) the two solitons collapse into each other in a distance $(\pi/4)e^{T/2}$ soliton units, this being one quarter-period of a periodic oscillation of the solitons in a bound state, where $T = t_2(0)-t_1(0)$ is the initial separation between the solitons at $x = 0$. In case (iii) the two solitons move apart in opposite directions, each soliton approaching a fixed terminal 'velocity'. (Note that in this 'quasiparticle' analogy the roles of time and distance are reversed from their conventional mechanical usage, so 'velocity' has physical units of time/distance, both being normalised in soliton units.)

In the more pertinent case here of an infinite train of solitons, the interaction of solitons in both directions in the pulse train has to be accounted for. The simplest hypothesis is that intersoliton forces add linearly on a nearest-neighbour approximation, justified by the exponential decrease of the intersoliton force in Equation 5.6 with increasing separation of the solitons. This turns out to be satisfactory in the case generalising conditions (ii) and (iii) above, when adjacent solitons have relative initial phases of 0 or π. Then the phases again remain constant throughout the motion and Equation 5.6a generalises to

$$d_x^2 q_k = 4e^{-T} \cos\alpha \left[e^{-(q_{k+1} - q_k)} - e^{-(q_k - q_{k-1})} \right] \tag{5.7}$$

with $(\alpha_{k+1} - \alpha_k) = \alpha = 0$ or π and $q_k = t_k - kT$ where T is the fundamental repetition period of the soliton pulse train. Equation 5.7 is known in the mathematical literature of nonlinear equations as the Toda lattice equation [7], and is completely integrable by the inverse scattering method. Simple stationary solutions of Equation 5.7 can be found by very elementary methods, however; the condition q_k

= constant = q_0 reduces the intersoliton forces to zero, and the lattice of displacements $\{q_k\}$ is in equilibrium.

5.3 Pulse position modulation

The situation described above, in which a periodic soliton train is perturbed by displacements in time $\{q_k\}$ of each soliton, is pulse position modulation (PPM). The fundamental limitations of *linear* PPM are the fibre dispersion, since the smallest pulse width which can be transmitted in the linear regime over a length x of fibre with dispersion D is $\tau = (2|D|x)^{1/2}$, and the fibre loss. The use of solitons, which are nonlinear phenomena, removes the dispersion limitation, and the use of erbium-doped fibre amplifiers (or Raman fibre amplifiers) removes the loss limitation. The first question to be addressed when studying a nonlinear dynamical system concerns the stability of its stationary states. This can be done easily in the case of Equation 5.7, simply by linearising the equation for small perturbations about the equilibrium [8]. Replacing the exponentials by their small-argument approximations gives

$$d_x^2 q_k = 4e^{-T}\cos\alpha\,(2q_k - q_{k+1} - q_{k-1}) \tag{5.8}$$

and solutions of this differential-difference equation are

$$q_k = \{Q\cos(\lambda x) + V\sin(\lambda x)\}e^{ik\beta} \tag{5.9}$$

where β is an arbitrary number in the range $-\pi \leq \beta \leq \pi$, and Q and V are constants determined from the initial conditions. The eigenvalue λ satisfies

$$-\lambda^2 = 8e^{-T}(1-\cos\beta)\cos\alpha \tag{5.10}$$

This result can be validated by a more rigorous full stability analysis [9]. More general real-valued solutions can be constructed by Fourier superposition of the basic modes of Equation 5.9 in the form

$$q_k = \tfrac{1}{2\pi}\int_{-\pi}^{\pi}\{Q(\beta)\cos(\lambda x) + V(\beta)\sin(\lambda x)\}e^{ik\beta}\,d\beta \tag{5.11}$$

The stability of the perturbations of position requires that λ be real, or alternatively λ^2 positive; if any λ^2 is negative, the corresponding mode of Equation 5.9 grows exponentially with distance. From Equation 5.10 stability only holds for $\cos\alpha < 0$, corresponding to the repulsive intersoliton phase $\alpha = \pi$ the other phase $\alpha = 0$ is unstable. The characteristic distance scale over which the instability exhibits its effects, which we call the *stability length* L, is

$$L = \min_{-\pi \leq \beta \leq \pi} |\lambda|^{-1} = \tfrac{1}{4} e^{T/2} \tag{5.12}$$

The stability length L is plotted in Figure 5.1 as a function of the soliton interval T for the specific fibre described after Equation 5.1, and summarised in Table 5.1, with soliton FWHM of 10 ps and 20 ps. In the example of soliton units quoted above, L is about 14400 km for 10 ps FWHM solitons spaced apart by 100 ps (soliton time unit $\tau = 5.7$ ps, $T = 17.6$ soliton units). If the soliton time unit is scaled by some factor η, the characteristic length L in physical units scales by η^2; thus, for 5 ps pulses spaced by 50 ps the stability length $L = 3600$ km. Due to the exponential dependence of the stability length on the pulse spacing T it decreases very rapidly for quite small reductions in T. Because the soliton length L_0 is inversely proportional to the dispersion parameter D, the stability length scales inversely with D.

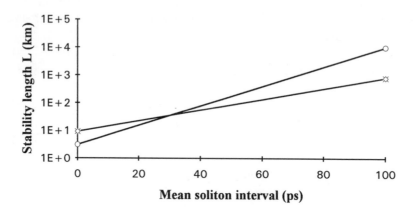

Figure 5.1 *Dependence of stability length on mean soliton period for a fibre with $D = -3.7$ ps² km⁻¹, effective area = 25 μm², (o -10 ps, * - 20 ps)*

Table 5.1 *Summary of fibre parameters*

Fibre dispersion parameter D	-3.7 ps² km⁻¹
Fibre attenuation α	0.2 dB km⁻¹
Effective area of fibre core	25 μm²
Soliton FWHM τ_s	10 ps
Soliton time unit τ	5.7 ps
Soliton length unit L_0	8.7 km
Amplifier spacing L_1	5 km
Optical carrier wavelength	1.55 μm

In the stable case, the dependence of λ on β in Equation 5.11 means that information impressed on the pulse train as PPM at $x = 0$ will disperse as the train progresses along x. For example, consider the case $Q(\beta) = q_0 = $ constant and $V(\beta) = 0$, for which at $x = 0$ $\partial_x q_k = 0$ and

$$q_k = 0 \qquad \text{for } k \neq 0$$
$$= q_0 \neq 0 \qquad \text{for } k = 0 \tag{5.13}$$

corresponding to a displacement of q_0 at the site $k = 0$ only, all other soliton positions being undisturbed from the equilibrium position of a periodic train. As x increases from zero, the term containing λx in Equation 5.11 becomes significant, and the displacements are no longer zero for sites with $k \neq 0$. The integral Equation 5.11 can be calculated explicitly for this case with the result that

$$q_k = q_0 J_{2k}(x/L) \tag{5.14}$$

where $J_n(z)$ represents the Bessel function of order n and argument z. Significant dispersion takes place over the same length scales as the instability demonstrated in the previous paragraph, Equation 5.12. In this simple example the displacement of the initially excited pulse at $k = 0$ falls to zero at a distance $x = 2.405\ L$, with the complete transfer of the modulation onto neighbouring pulses. Thus it is seen that the limiting distance scale for *small-deviation* PPM is given by Equation 5.12. For distances significantly less than the stability length ($x/L \ll 1$) Equation 5.14 describes essentially nearest-neighbour interpulse interference, because the Bessel function decays very rapidly with k if x/L is small. In this approximation the interference present on the pulse position at site k due to its neighbours on sites $k\pm1$ is obtained by superposition of terms from Equation 5.14 as

$$e_k \approx q_{k-1} J_{-2}(x/L) + q_{k+1} J_2(x/L).$$

If the mean-square modulation on each soliton is $\sigma_q^2 = \langle q_k^2 \rangle$ and the $\{q_k\}$ are uncorrelated then this leads, after squaring, averaging and approximating the Bessel functions by the dominant terms of their power series, to the approximation

$$\sigma_e^2 \approx 2\sigma_q^2 J_2^2(x/L) \approx \sigma_q^2 (x/L)^4/32 \tag{5.15}$$

for the mean-square interpulse interference $\sigma_e^2 = \langle e_k^2 \rangle$. For $x/L = 0.1$, Equation 5.15 gives the ratio $\sigma_q^2/\sigma_e^2 = 2^5 \times 10^4 = 55$ dB. There is also a reduction of the mean-square *signal* component of the position of the pulse on each site by a factor $J_0^2(x/L)$; this is close to unity if x/L is small. The approximate result from Equation 5.15 is unchanged if the unstable pulse train (with attractive interactions) is used rather than the stable one; in the unstable case the Bessel functions in Equation

5.14 are replaced by modified Bessel functions $I_{2k}(x/L)$, with the same leading order approximation for I_2 as for J_2.

It will transpire that large-deviation PPM is limited to shorter length scales for the same pulse period T, because larger pulse deviations from the nominal equilibrium of a periodic pulse train cause larger interpulse interaction forces, which evolve dynamically over shorter length scales. Thus, from the point of view of maximising the distance over which a given bandwidth can be transmitted, small-deviation PPM is advantageous. In a communication channel this conflicts with the requirement to maximise the analogue output signal-to-noise ratio, which in a linear channel would require as much as possible of the available sample period T to be used for the deviation of the pulse position. However, a soliton pulse requires a significant amount of energy, by definition much larger than a linear pulse since the nonlinear refractive index of the fibre must be exploited to a significant extent, which requires high peak optical intensities in the fibre medium.

The principal source of noise in long fibre communication systems is expected to be amplified spontaneous emission (ASE) from the erbium-doped fibre amplifiers used to compensate for fibre loss [10]. Each amplifier of power gain G is sufficient to compensate for the linear loss of the fibre between itself and the preceding amplifier, and introduces white noise of power spectral density $(G-1)h\nu$ (W Hz^{-1}) into each fibre section between amplifiers [10]. If the fibre loss is γ, the inter-amplifier spacing is L_1 so that $G = \exp(\gamma L_1)$, and the required span of the fibre link is x, then $N = x/L_1$ amplifiers are needed, and noise of spectral density

$$S = N(G\text{-}1)h\nu \qquad\qquad\qquad (5.16)$$

is added to the fibre output, where ν is the optical carrier frequency and h is Planck's constant. Part of this noise energy is worked into the pulse by the nonlinearity of the fibre medium as fluctuations of its velocity, and hence its position (the Gordon-Haus effect [10]), in addition to its appearance as additive amplitude noise. The ratio of noise spectral density S to pulse energy E is generally small ($S/E \ll 1$), increasing linearly with x according to Equation 5.16.

ASE generates fluctuations in the velocities of solitons leaving the optical amplifiers. It has been shown by Gordon and Haus [10] that if each amplifier adds a velocity whose mean square value is $\sigma_p{}^2$, then the normalised mean-square fluctuation in arrival time of an isolated soliton arriving at the position x along the fibre is

$$\sigma_v^2 = \frac{\sigma_p^2}{3L_1}x^3 \qquad\qquad\qquad (5.17)$$

where L_1 is the normalised length between amplifiers and

$$\sigma_p^2 = \frac{2}{3}\frac{h\nu}{E}(G-1) \qquad\qquad\qquad (5.18)$$

where $E = 2P\tau$ is the energy of the soliton pulse. For a fibre with the parameters quoted previously, summarised in Table 5.1, these authors have shown that the RMS fluctuation of position is approximately equal to one soliton unit at a distance of 2360 km for a soliton of 10 ps FWHM. This distance is 0.16 of the stability length (14400 km) in this case. The expression in Equation 5.17, which was obtained in [10] only for the case of an isolated 1-soliton, requires $x \ll L$ when the intersoliton interaction is also accounted for [11], as will be shown in the next paragraph. Thus the values calculated by Gordon and Haus are well within the limits of validity of Equation 5.17. This fluctuation of soliton arrival time of one soliton unit is comparable with, or larger than, the pulse deviation impressed as modulation in the case of small-deviation PPM, and therefore implies a very low effective SNR for the soliton PPM, even though the classical SNR E/S may be quite high.

Because of the exponential dependence of the stability length on the intersoliton period T (Equation 5.12), a modest reduction in T will increase x/L to larger values of order 1 or greater. In this regime the effectiveness of the channel is affected significantly by a combination of dispersion (Equation 5.15) and GH noise. A more exact calculation of the GH noise is required here, due to the failure of approximations leading to Equation 5.16. This can also be obtained from the linearised theory leading to Equation 5.11. If, instead of the initial condition in Equation 5.13, we choose instead

$$
\begin{aligned}
q_k &= 0 && \text{for all } k \\
d_x q_k|_{x=0} &= 0 && \text{for } k \neq 0 \\
d_x q_k|_{x=0} &= p_0 && \text{for } k = 0,
\end{aligned}
\tag{5.19}
$$

corresponding to all solitons at the equilibrium position at $x = 0$ with a velocity of p_0 imparted to the soliton at $k = 0$, then the solution of the integrals resulting in Equation 5.11 is

$$
q_k(x) = p_0 \int_0^x J_{2k}(x' / L) dx'
\tag{5.20}
$$

For initial conditions corresponding to velocities p_k added to each soliton at a point $x = x'$ along the fibre

$$
\begin{aligned}
q_k|_{x=x'} &= 0 && \text{for all } k \\
d_x q_k|_{x=x'} &= p_k(x')
\end{aligned}
\tag{5.21}
$$

the result is obtained by superposition of elementary solutions of the form of Equation 5.20 to obtain

$$q_k(x) = \sum_{r \in Z} p_r(x') \int_0^{x-x'} J_{2(k-r)}(x''/L)dx''$$

For a continuous distribution of sources of velocity perturbations along the fibre length, corresponding to the output of each fibre amplifier, then

$$q_k(x) = L_1^{-1} \sum_{r \in Z} \int_0^x p_r(x') \int_0^{x-x'} J_{2(k-r)}(x''/L)dx''dx'$$

where L_1 is the (normalised) spacing between amplifiers (here we replace a sum over discrete amplifier positions by an integral over a continuum with the same 'density' of amplifiers L_1^{-1}. This approximation is valid as long as $L_1 \ll L$). Finally, statistical averages of the last expression can be taken, assuming that all the $p_r(x')$ are independent with mean square σ_p^2, to obtain the final result

$$\sigma_v^2 = \langle q_k^2(x) \rangle = L_1^{-1} \sigma_p^2 \sum_{r \in Z} \int_0^x \left\{ \int_0^{x-x'} J_{2(k-r)}(x''/L)dx'' \right\}^2 dx' \qquad (5.22)$$

When the range x is much less than the stability length L the Bessel functions can be approximated in Equation 5.22 by $J_0 = 1$ and $J_{2k} = 0$ for $k \neq 0$, and the approximate result in Equation 5.17 is obtained. Figure 5.2 illustrates the difference in large x/L behaviour of the GH noise for the 1-soliton and soliton train case.

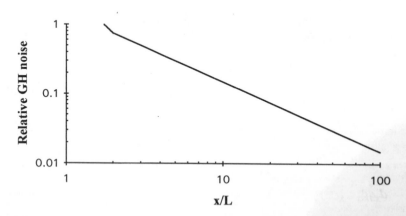

Figure 5.2 *Reduction of Gordon-Haus noise in an interacting soliton train relative to its behaviour in an isolated soliton*

A particular attraction of small-deviation PPM is that it can be demodulated by simple linear filters with bandwidths no larger than the required signal bandwidth [12]. This incurs a quite considerable penalty in SNR over ideal matched-filter detection, but there is also a significant gain in the reduced complexity of receiver design and realisation. The SNR penalty cannot, however, be offset by increasing the pulse energy, as would be the case in a linear system. Increasing the energy of a soliton by the factor η also requires the reduction of its pulse width by the same factor; this increases the pulse bandwidth by η, but all the additional energy is lost when the pulse stream is subjected to the detector filter over the analogue signal bandwidth since it falls outside the filter passband.

5.4   The analogue soliton PPM communications channel

It is possible to study the behaviour of the soliton PPM channel under conditions appropriate to the transmission of a strictly analogue signal such as analogue video. However, current trends in the development of wideband communications networks are heavily weighted towards *digital* modulation as the format in which these signals will be represented; an ultra-wideband *analogue* network is not envisaged. Nevertheless, the performance of a digital transmission system often depends critically on features of the channel which are purely analogue in nature (its bandwidth, rise-time, carrier-to-noise ratio, etc.). From this viewpoint, the only factor which distinguishes digital from analogue modulation is the coding gain of the former, which permits low error rates to be achieved at channel SNRs which would make analogue transmission impossible, at the cost of bandwidth and receiver complexity. PPM acts by periodically sampling an analogue waveform, and transmitting the samples encoded as the positions of the pulses; it is immaterial whether the sampled waveform is itself digitally encoded, as for example is the bandlimited signal given by

$$f(t) = \sum_{k \in Z} a_k \frac{\sin\{\Omega(t - kT)\}}{\Omega(t - kT)} \qquad (5.23)$$

where $\Omega = \pi/T$, and $a_k \in \{0, 1\}$ are binary-valued coefficients. In the special case of uniformly-sampled PPM, where the PPM sampling takes place at the times $t = kT$, then the PPM transfers the digital values a_k directly to the encoded pulse positions, a case we refer to as *direct digital PPM*. Any other sampling scheme leads to *indirect digital PPM*, in which the underlying signal may be digital, but the PPM behaves in an essentially analogue manner.

In order to detect *direct digital PPM* with BER better than 10^{-9}, it is necessary to achieve RMS pulse jitter less than 1/3 of the difference in times between the two times representing the 0 and 1 symbols. If σ is the RMS pulse jitter, assumed to be Gaussian, and the two signal times are $\pm q_0$, then this implies $q_0 > 6\sigma$ (or equivalently, $(q_0/\sigma)^2 > 15.6$ dB). (This follows from the well-known formula $P_e = \text{erfc}(q_0/\sigma)$ for the BER P_e.). Assuming that the solitons are spaced far enough apart to be noninteracting, so that the simple approximation of Equation 5.17 can be used, this would require the RMS Gordon-Haus noise to be 0.16 soliton units for a deviation of $q_0 = 1$ soliton unit, which is achieved at a distance of $x = 164$ soliton units = 1430 km using the standard fibre parameters of Table 5.1 in Equation 5.17. Maintaining the interpulse interference σ_e^2 less than the GH noise requires, from Equation 5.15, $x/L < 0.97$. From Equation 5.12 this implies $T > 13$ soliton units = 74 ps, which would permit a digital capacity of 13.5 Gb s^{-1}. It is possible using soliton PPM at moderate values of x/L (actually $x/L < \pi/2$) to compensate for the effects of intersoliton interference using discrete-time transversal filters in the receiver [14]; use of such compensation at the limiting value of $x/L = \pi/2$ would increase the capacity slightly to 14.5 Gb s^{-1}. Having computed the required value of x/L, a correction can be made for the true value of GH noise using the more accurate Equation 5.22, but this correction is small in this case and does not significantly alter the calculations, which are in any case only estimates. These calculations show that very large capacities can be achieved by small-deviation direct digital soliton PPM, over distances which are typical of long-haul terrestrial networks in the UK and Europe. By comparison, for ASK modulation the theory of Gordon-Haus predicts that the product of capacity and distance is constant; for the standard fibre parameters in Table 5.1 this constant is 23600 Gb s^{-1} km at BER = 10^{-9}. Thus, for ASK modulation, the theory predicts a capacity of 16.5 Gb s^{-1} at a distance of 1430 km. However, soliton interactions would reduce this capacity somewhat, although the reduction is difficult to estimate precisely because we do not at present have a fully developed analytical theory for ASK modulation. Thus we conclude that the PPM capacity is at least comparable with the ASK capacity over the same distance of 1430 km.

Indirect digital soliton PPM, in which samples are taken to represent a digitally encoded waveform like Equation 5.23, behaves in exactly the same way as the direct form of the modulation, with the sole difference that the digital detector placed after an analogue PPM demodulator requires a more sophisticated synchronisation system, since the digital waveform and the PPM sampling do not necessarily share a common time reference. In particular, the same ratio of pulse deviation to pulse jitter must be achieved for indirect PPM as for direct PPM.

A significant advantage of PPM is that it can be used for true time-division multiplex transmission (TDM). The classical definition of TDM is the allocation of the full capacity of an analogue channel to each of a number of signal sources in sequence for a predetermined interval of time. In the duration of each such time interval, one PPM soliton can be transmitted to represent a sample of each of the signal sources in turn, resulting in the interleaving in time of the source channels on the multiplexed PPM pulse train. The analogue nature of this multiplex arrangement effectively eliminates any problems in transmission due to asynchronicity of the various source channels, which may run at arbitrary bit-rates consistent with the Nyquist rate for proper sampling of the channel. Thus, a PPM pulse train whose mean pulse rate is v can interleave M sub-channels, each sampled at v/M, if the 2-sided bandwidth of each sub-channel is less than v/M, each with a capacity of up to v/M (b s^{-1}) if encoded according to Equation 5.23.

5.5 Nonlinear PPM phenomena

In PPM over a *linear* channel where the pulse width is much less than the pulse repetition time ($T \gg 1$) it is possible to increase the post-demodulator SNR for a fixed receive-end SNR by increasing the deviation of the modulation to use more of the available time between pulses. This leads to the opposite limit to that considered above, of *large-deviation* soliton PPM, which can also be treated qualitatively. Suppose that, having designed a small-deviation PPM link as outlined above, we wish to increase its SNR by increasing the deviation to some value $\pm mT/2$, where $0 < m < 1$ is the modulation index. The minimum separation between successive solitons will then be $(1-m)T$ and for those solitons separated by this minimum time the nonlinear interaction between them will be enhanced due to the exponential dependence of the 'force' in Equation 5.7. This means that their positions will evolve on shorter distance scales than the stability length (Equation 5.12), with a more rapid loss of information to neighbouring solitons than in the small-deviation case. To mitigate this increased interaction, the mean pulse period T must be increased until the new value T' is such that $(1-m)T' = T$, restoring the worst-case intersoliton separation to that of the small-deviation case. This reduces the transmission capacity by a factor $(1-m)$. In addition, large-deviation analogue PPM requires more sophisticated detection methods in order to avoid distorted reproduction, increasing the cost of the receiver [12].

Mathematical analysis of the solutions of Equation 5.7 reveals some spectacular effects if the modulation index m is increased without reduction of the mean pulse rate. As an example, consider the initial condition

$$q_k(0) = \ln\left\{\frac{\cosh\left(\frac{\pi_0}{2}\right) + \cosh\left(\frac{k\tau_0}{2}\right)}{\cosh\left(\frac{\tau_0}{2}\right) + \cosh\left(\frac{(k+1)\tau_0}{2}\right)}\right\} \tag{5.24}$$

This represents a smooth transition of the displacements q_k between constant limiting values $q_k \to \tau_0/2$ as $k \to -\infty$ and $q_k \to -\tau_0/2$ as $k \to \infty$, with τ_0 being the maximum deviation of the displacement. For sufficiently large τ_0 the change in displacement is concentrated almost entirely on a few solitons near $k = 0$. The reason for choosing this particular initial condition is that it represents the 2-soliton solution of the Toda lattice equation [7] (Equation 5.7 with the upper sign representing a repulsive intersoliton force), evolving with x into

$$q_k(x) = \ln\left\{\frac{\cosh\left(\frac{\pi_0}{2}\right)\cosh(X)+\cosh\left(\frac{k\tau_0}{2}\right)}{\cosh\left(\frac{\tau_0}{2}\right)\cosh(X)+\cosh\left(\frac{(k+1)\tau_0}{2}\right)}\right\} \tag{5.25}$$

where $X = 4\beta x e^{-T/2}$ and $\beta = \sinh(\tau_0/4)$. For $X > 1$ Equation 5.25 evolves rapidly with x into two separate Toda lattice solitons propagating in opposite directions through the lattice of q_k- values [7]. (It is important here to distinguish between solitons of the NLSE and solitons of the Toda lattice, which are step-like distributions of the positions of the NLSE solitons).

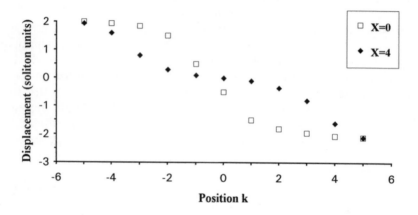

Figure 5.3 *Formation of two counter-propagating lattice solitons from the initial condition of equation (24): T = 16; $\tau_0 = 4$*

Figure 5.3 above illustrates the Toda lattice soliton phenomenon in the case $T = 16$, $\tau_0 = 4$ (normalised soliton time units). Interpreted in communications terms, any section of a sampled analogue waveform looking similar to Equation 5.24 at x = 0 will collapse rapidly under propagation of its PPM representation, the collapse taking place by emission of two lattice solitons (step-like disturbances which move rapidly in opposite directions through the pulse train), completely disintegrating the whole modulation pattern in the process. In the example of Figure 5.3 the initiation of the two Toda lattice solitons can be seen in the two step-like disturbances forming on either side of the symmetry axis $k = -1/2$; the modulation on the pulses $k = -2, -1, 0$ and 1 has already collapsed to near the equilibrium value $q_k = 0$ after a normalised propagation distance $X = 4$. Waveforms locally similar to Equation 5.24 near $k = 0$ with arbitrary τ_0 are not particularly pathological, and may be expected to occur with reasonably high probability in analogue waveforms. Such distributions will also occur in the *M*-ary digital soliton PPM proposed recently

[13]. Whatever the origin of such features may be, they will have a disastrous effect on the integrity of the reconstructed waveform after transmission of the PPM-coded signal; the characteristic distance scale over which this collapse takes place is $X \sim 1$, which translates into

$$x \sim L \, \mathrm{csch}(\tau_0/4) \tag{5.26}$$

For large modulation index (say, $m = 1/2$, $\tau_0 = 8$, $T = 16$) this distance is significantly less than the stability length L which was introduced in the small-deviation analysis considered earlier. The velocity of these disturbances is very high; once established ($X > 1$), each Toda soliton moves in time by one lattice spacing in a distance

$$\Delta x = L(\tau_0/4)\mathrm{csch}(\tau_0/4) \tag{5.27}$$

and by one soliton time unit (τ seconds) in $(1/T)$ times this quantity.

A full analysis of the Toda-lattice phenomena described in the previous paragraph would involve more sophisticated mathematical techniques. However, although this is only an example, it does illustrate that significant signal degradation will occur in the large-deviation case if the mean pulse period is not increased to compensate for an increase in modulation index in an attempt to increase SNR. It is also a very graphic example of the interesting dynamical effects that can occur in modulated soliton trains.

5.6 Conclusions

An analysis has been presented of the dynamical behaviour of soliton communications channels over long lengths at high transmission rates. The principal concern has been to demonstrate the utility of analogue concepts in performing analysis on this nonlinear channel, although analogue concepts can be equally well applied to digital modulation. The concepts of stability length and dispersive stability of linearised modulation have been introduced and applied to the cases of PPM. A new formula for the effects of ASE noise in the presence of significant pulse interaction has been derived, generalising a formula of Gordon-Haus from the 1-soliton case to that of an infinite train of interacting solitons. A design procedure has been sketched for PPM analogue modulation of solitons, including the effects of intersoliton interaction and amplified spontaneous emission noise, which appears both as additive channel noise and as fluctuations of the positions of the pulses. A representative PPM channel used to transmit a digitally encoded waveform performs at least as well as its conventional ASK counterpart; TDM networks can be envisaged using PPM.

An exhaustive theory of this channel has not been attempted, and indeed this requires much more development of mathematical tools before it could be achieved. The theory has been confined to exhibiting a few numerical examples which can be compared directly with published calculations [10] for digital ASK over the same fibre. Nevertheless, it is clear that the analogue behaviour of the soliton channel presents new and interesting dynamical problems, the solutions of which are also applicable to digital modulation.

5.7 References

1   Mollenauer, L. F., Evangelides, S. G. and Haus, H. A.: 'Long distance soliton propagation using lumped amplifiers and dispersion shifted fiber', *J. Lightwave Techol.*, 1991, **9**, pp. 194-197

2   Mollenauer, L. F., Neubelt, M. J., Evangelides, S. G., Gordon, J.P., Simpson, J. R and Cohen, L. G.: 'Experimental study of soliton transmission over more than 10000 km in dispersion-shifted fiber', *Opt. Lett.*, 1990, **15**, pp. 1203-1205

3   Mollenauer, L. F., Nyman, B. M., Neubelt, M. J., Raybon, G. and Evangelides, S. G.: 'Demonstration of soliton transmission at 2.4 Gbit/s over 12000 km', *Electron. Lett.*, 1991, **27**, pp. 178-179

4   Gordon, J. P.: 'Interaction forces among optical fibre solitons', *Opt. Lett.*, 1983, **8**, pp. 596-598

5   Butcher, P. and Cotter, D.: 'The elements of nonlinear optics' (Cambridge University Press, Cambridge, 1990)

6   Kodama, Y.: 'Optical solitons in a monomode fibre', *J. Stat. Phys.*, 1985, **39**, pp. 597-614

7   Toda, M.: Chap. 4 in 'Solitons', Eds. Bullough, R. K. and Caudrey, P. J., (Springer Topics in Modern Physics **17**, Springer-Verlag, Berlin, 1981)

8   Arnold, J. M.: 'Qualitative dynamics of modulated soliton pulse trains', Technical Digest 'Nonlinear guided wave phenomena', **15** (OSA, Washington DC, 1991), pp. 10-13

9   Arnold, J. M.: 'Stability theory for periodic pulse train solutions of the nonlinear Schrodinger equation', *IMA J. App. Math.*, 1994, **52**, pp. 123-140

10   Gordon, J. P. and Haus, H. A.: 'Random walk of coherently amplified solitons in optical fibre transmission', *Opt. Lett.*, 1986, **11**, pp. 665-667

11   Arnold, J. M.: 'Modulation and noise in soliton pulse trains', in 'Ultra-wideband short-pulse electromagnetics', Eds. Bertoni, H. L. , Carin, L. and Felsen, L. B. (Plenum Press, New York, 1993), pp. 267-274

12   Rowe, H. E.: 'Signals and noise in communication systems' (van Nostrand-Reinhold, Princeton, 1965)

13   Cryan, R. A., Unwin, R. T., Massarella, A. J. and Sibley, M. J. N.: 'Digital PPM employing temporal solitons', IEE Colloquium 'Nonlinear effects in optical fibres', Digest 1990/159, 14/1-5, 1990

14   Arnold, J. M.: 'Saturated growth of Gordon-Haus noise in interacting soliton pulse trains', Technical Digest 'Nonlinear guided wave phenomena', **15** (OSA, Washington DC, 1993), pp.176-181

Chapter 6

Optical fibre digital pulse position modulation

R.A. Cryan, R.T. Unwin and J.M.H. Elmirghani

6.1 Introduction

Continuous pulse time modulation (PTM) techniques have been proposed for the economic short-haul distribution of analogue signals over optical fibres. The scenarios envisaged are the transmission of high quality video [1-3] or alternatively control and instrumentation signals [4,5] over optical fibre employing LED transmitters and PIN photodiode detectors. This type of modulation uses a binary pulse amplitude, the analogue signal being conveyed by the continuously varying pulse edge. The PTM schemes have the advantage that the analogue signal can be transmitted at a reduced bandwidth than that required by pulse code modulation (PCM). This allows the use of inexpensive, low bandwidth optical fibres which when combined with the LED transmitter and PIN detectors leads to an economical system.

Recently, a discrete PTM scheme has been suggested for long-haul point-to-point links over single-mode fibre. The scheme under consideration utilises the pulse position modulation (PPM) format, whereby M bits of PCM are conveyed by positioning a single high energy pulse in one of 2^M positions. In contrast to its analogue equivalent, digital PPM actually consumes more bandwidth than that required by PCM. However, even the most advanced of today's optical fibre communications systems utilises no more than 0.1 % of the 20 THz of bandwidth available in single-mode fibre, so it would seem reasonable to use this form of discrete PTM as a means of improving receiver sensitivity. Calculations have shown that digital PPM can offer improvements of 5-11 dB [6-14] (depending on the coding level, bandwidth and detection technique) in receiver sensitivity when compared with PCM. This represents an increase in point-to-point transmission of 25-55 km. An experimental digital PPM system has been shown to offer an improvement of 4.2 dB over PCM [15].

The pre-detection filter for this discrete PTM format is extremely complicated when compared to that of PCM (normally a third-order Butterworth filter). It consists of a pre-whitened matched filter in cascade with a proportional-derivative-delay (PDD) network. In the experimental systems [15,16], a complex algorithm was used to determine the component values of a 10th order ladder network which was matched to the incident pulse shape. The PDD network was realised using coaxial delay lines, buffers, switched attenuators and a summing amplifier. The attenuator settings were dependent on the number of PPM slots and the pulse width. In this form, the pre-detection filter would be unrealistic in a commercial system.

The first part of this chapter investigates the penalty incurred in receiver sensitivity by adopting sub-optimal filtering such that receiver complexity can be significantly simplified. A 140 Mbit/s, 1.3 μm digital PPM system employing a transimpedance CC-CE PIN-BJT pre-amplifier is considered. Original sensitivity calculations have been performed for four pre-detection filters: the optimum filter, a matched filter, a third-order network where the pole locations have been optimised and finally a third-order Butterworth filter. The maximum sensitivities offered by these systems are -47.4 dBm, -47 dBm, -46.5 dBm and -46.3 dBm compared to -38.8 dBm for a PCM system. The sub-optimum systems offer sensitivity improvements of at least 7.5 dB over PCM and are within 1.1 dB of the optimum. Hence, receiver complexity can be significantly simplified without incurring enormous penalties in receiver senstivities. In particular, a simple Butterworth filter can be employed with only a 1.1 dB degradation in receiver sensitivity.

The PPM performance in [6-16] has been analysed on the assumption of perfect transmitter-receiver synchronisation. Practically the timing information has to be extracted at the receiver from the PPM data stream. PPM synchronisation on the slot and frame levels has previously been considered [17-21]. The extracted slot and frame clocks can be impaired by a variety of sources including the channel imperfections and the receiver noise. The optical fibre channel imperfections and the receiver noise will be common to the PPM system and the other systems such as PCM. Therefore it is of interest to establish the performance bound due to the PPM format inherent limitations, namely the impact of the PPM pattern-dependent (systematic) jitter on the slot and frame synchronisers and hence the system error performance.

The second part of this chapter presents an original spectral characterisation for the PPM format showing the presence of discrete slot and frame rate components together with the signalling noise or the continuum. The impact of the PPM self-noise (continuum) on the slot timing variance is evaluated for various PPM orders and phase-lock loop bandwidths. The resultant wrong-slot error probability due to the slot clock jitter is evaluated in terms of the system parameters and shown to be optimisable. Frame synchronisation is analysed for an original class of synchronisers that employs natural occuring sequences [19]. The frame timing variance is evaluated as a function of the phase-lock loop bandwidth and the number of natural sequences tracked. The system wrong-slot errors due to frame clock jitter are evaluated and the analysis enables the identification of the optimum PPM parameters for a given required error performance when slot and frame clock impairments are taken into consideration.

6.2 Analysis

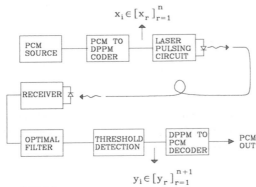

Figure 6.1 *The digital PPM system*

A block diagram of the physical system is shown in Figure 6.1. M bits of binary PCM are coded into a set of symbols $x_i \in [x_r]_{r=1}^n$ where $n = 2^M$. As illustrated in Figure 6.2, each digital PPM symbol is formed by subdividing the symbol interval T_n termed the frame time, into n time slots each of width wT_n such that $nw = m < 1$. We call m the modulation index and leave a guard interval of $(1-m)T_n$ at the end of each frame for timing extraction purposes and to prevent inter-frame interference due to dispersion. The voltage $v_o(t)$ at the output of the receiver is passed through a pre-detection filter in order to optimise the signal-to-noise ratio before being presented to the detection device.

The output voltage crosses a threshold level v_d, with positive slope, at time t_d. A comparison of t_d with a clock signal leads to an estimate $y_j \in [y_r]_{r=1}^n$ consisting of the n pulse positions plus erasures of the symbol x_i. For a slightly dispersive optical fibre channel the estimate y_j may differ from the true symbol x_i due to three main sources of error.

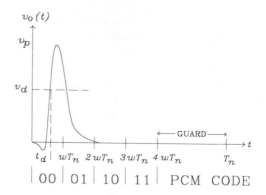

Figure 6.2 *The digital PPM signal*

6.2.1 Erasures

An erasure error occurs whenever noise destroys the pulse thus preventing detection. The probability of an erasure error P_r, assuming the receiver output noise voltage is a Gaussian random variable, is [6,22]:

$$P_r = 0.5\mathrm{erfc}\left(\frac{Q_r}{\sqrt{2}}\right) \tag{6.1}$$

where

$$Q_r^2 = \frac{\left(v_p - v_d\right)^2}{\left\langle n_o(t)^2 \right\rangle} \tag{6.2}$$

and where $\left\langle n_o(t)^2 \right\rangle$ is the mean square receiver output noise, v_d is the receiver output at the threshold crossing time t_d and v_p is the peak receiver output which occurs at time t_p.

6.2.2 Wrong- slot errors

These occur when noise on the leading edge of the pulse produces a threshold crossing in the time slot immediately preceding or following that containing the pulse. The probability of a wrong slot error P_S is given by

$$P_s = \mathrm{erfc}\left(\frac{Q_s}{\sqrt{2}}\right) \tag{6.3}$$

where

$$Q_s^2 = \left(\frac{mT_n}{2n}\right)^2 \frac{1}{\langle n_o(t)^2\rangle}\left(\left.\frac{dv_o}{dt}\right|_{t_d}\right)^2 \tag{6.4}$$

6.2.3 False alarm errors

In the interval between the start of the frame and the arrival of the signal pulse, the receiver output voltage may cross the threshold due to noise with probability

$$P_f = \left(\frac{mT_n}{n\tau_r}\right)\mathrm{erfc}\left(\frac{Q_r}{\sqrt{2}}\right) \tag{6.5}$$

where

$$Q_f^2 = \frac{v_d^2}{\langle n_o(t)^2\rangle} \tag{6.6}$$

The number of uncorrelated samples per time slot can be estimated in terms of the time τ_r at which the autocorrelation function of the receiver filter has become small, as $(mT_n/n\tau_r)$. The probability per time slot of a false alarm error is then approximated by $P_f = (mT_n/n\tau_r)P_t$ when $P_f \ll 1$.

6.3 Performance criterion

In order to see how the error sources affect system performance in comparison to PCM it is convenient to compare the two systems with respect to the average binary error probability P_{eb}. We define the performance criterion as

$$P_{eb} = P_e = 10^{-9} \tag{6.7}$$

where P_e is the error probability of an equivalent PCM and P_{eb} is derived below. The average error probability in a digital PPM system with n different symbols, assuming that all symbols are equally probable, is given by

$$P_{es} = \frac{1}{n} \sum_{k=1}^{n} P_s(k) \tag{6.8}$$

where $P_s(k)$ is the probability that $y_j \neq x_i$. For a pulse in slot k, assuming that the joint probability of an erasure and a false alarm is negligible

$$P(0|k) = P_r \qquad \text{(an erasure)}$$

$$P(1|k), \ldots, P(k-2|k) = P_f \qquad \text{(false alarm)}$$

$$P(k-1|k) = P_f + \frac{P_s}{2}$$

$$P(k|k) = 1 - P_s - P_r - (k-1)P_f \quad \text{(correct)}$$

$$P(k+1|k) = \frac{P_s}{2}$$

$$P(k+2|k), \ldots, P(n|k) = 0$$

hence $P_s(k)$ becomes

$$P_s(k) = P_r + P_s + (k-1)P_f \tag{6.9}$$

Substituting Equation 6.9 into 6.8 and evaluating gives

$$P_{es} = P_r + P_s + \frac{(n-1)}{2} P_f \tag{6.10}$$

from which the average binary error probability P_{eb} can be determined [23]

$$P_{eb} = \frac{n}{2(n-1)} P_{es} = \frac{n}{2(n-1)} [P_r + P_s] + \frac{n}{4} P_f \tag{6.11}$$

6.4 The optimal filter

Garrett [6] has shown that the optimum filter for estimating the pulse arrival time in a direct detection digital PPM system is a pre-whitened filter, matched to the incident optical pulse shape, in cascade with a proportional-derivative-delay (PDD) network as illustrated in Figure 6.3. The transfer function of the optimal filter is given by

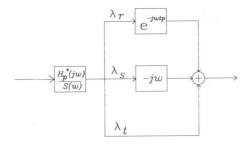

Figure 6.3 *The optimum pre-detection filter*

$$G(j\omega) = \frac{H_p^*(j\omega)}{S(\omega)} \cdot \left[\lambda_f - j\omega\lambda_s + \lambda_r \cdot e^{-j\omega t_p} \right]$$ (6.12)

where $S(\omega)$ is the noise power spectral density.

The Lagrangian multipliers λ_f, λ_s and λ_r are factors which are to be determined in terms of the system parameters. $G(\omega)$ contains an arbitrary scalar factor, thus it is possible to write

$$\lambda_f + \lambda_s + \lambda_r = 1$$ (6.13)

Hence there are two independent multipliers to be determined. The PDD network is used to minimise the probability of occurrence of the three possible errors by shaping the pulse before detection. The λ_f weighting is used to counter false alarm errors by increasing the proportional content of the original signal, thus increasing the pulse amplitude. Wrong-slot errors are reduced by sharpening the detection edge of the pulse. This can be achieved by increasing the λ_s weighting in the derivative branch. Finally, by altering the pulse delay time with λ_r and adjusting λ_f and λ_s, we can minimise erasure errors by lowering the threshold point of detection. The pulse energy required for correct detection, $P_e = 10^{-9}$, may be minimised for various operating conditions (e.g. high bandwidth, high n) by an optimum choice of the λ weightings.

6.5 Algorithm for determining the receiver sensitivity

In this section, the algorithm for determining the receiver sensitivity for a digital PPM system is described. By convolving the received optical pulse with the pre-amplifier impulse response and the impulse response of the pre-detection filter an expression is derived for the pulse shape presented to the threshold detector. Knowledge of this pulse shape along with the mean-square noise allows the individual error probabilities to be determined and therefore system performance. Four pre-detection filters are considered.

6.5.1 Optimal filter output

For a received pulse energy b and pulse shape $h_p(t)$ such that $\int_{-\infty}^{\infty} h_p(t)dt = 1$, the pre-amplifier output voltage is:

$$v_{pa}(t) = bRz_t(t) * h_p(t) = \frac{bR}{2\pi} \int_{-\infty}^{\infty} Z_t(j\omega)H_p(j\omega) \cdot e^{j\omega t} d\omega \tag{6.14}$$

where R is the photodiode responsivity and $Z_t(j\omega)$ is the pre-amplifier frequency response. The output of the optimal filter may now be expressed as:

$$
\begin{aligned}
v_o(t) &= v_{pa}(t) * y_n(t) * g(t) \\
&= \frac{bR}{2\pi} \int_{-\infty}^{\infty} |H_p(j\omega)|^2 Y_n(j\omega)Z_t(j\omega)\left[\lambda_f - j\omega\lambda_s + \lambda_r e^{-j\omega t_p}\right] \cdot e^{j\omega t} \, d\omega
\end{aligned}
\tag{6.15}
$$

where $y_n(t)$ is the impulse response of the noise whitening filter and $g(t)$ is the impulse response of the optimal pre-detection filter.

Defining

$$I_k(t) = \frac{1}{2p} \int_{-\infty}^{\infty} (j\omega)^k |H_p(j\omega)|^2 Y_n(j\omega)Z_t(j\omega) \cdot e^{j\omega t} d\omega \tag{6.16}$$

allows $v_o(t)$ to be expressed as

$$v_o(t) = bR[\lambda_f I_0(t) - \lambda_s I_1(t) + \lambda_r I_0(t - t_p)] \tag{6.17}$$

Letting

$$J_k = I_k(0) \quad \text{and} \quad K_k = I_k(t_p) \tag{6.18}$$

which are definite integrals allows v_p, v_d and v_d' to be expressed as

$$v_d = v_o(t_d) = bR\left[\lambda_r K_0 - \lambda_s J_1 + \lambda_f J_0\right] \tag{6.19}$$

$$v_p = v_o(t_p) = bR\left[\lambda_r J_0 - \lambda_s K_1 + \lambda_f K_0\right] \tag{6.20}$$

$$v'_d = \left. \frac{dv_o(t)}{dt} \right|_{t_d} = bR\left[-\lambda_r K_1 - \lambda_s J_2 + \lambda_f J_1\right] \tag{6.21}$$

The mean-square noise at the output of the optimal pre-detection filter is given by

$$\left\langle n_o(t)^2 \right\rangle = \frac{1}{2\pi} \int_{-\infty}^{\infty} S(\omega) \left| H_p^*(j\omega) Z_t(j\omega) Y_n(j\omega) \left[\lambda_f - j\omega\lambda_s + \lambda_r \cdot e^{j\omega t_p} \right] \right|^2 d\omega \tag{6.22}$$

Since a noise-whitening filter is employed the noise power spectral density can be taken to be white over the frequency range of interest and Equation 6.22 may be written as:

$$\left\langle n_o(t)^2 \right\rangle = S_o\left[-\lambda_s^2 L_2 + \left(\lambda_r^2 + \lambda_f^2\right)L_0 + 2\lambda_f\lambda_r M_0 - 2\lambda_r\lambda_s M_1\right] \tag{6.23}$$

where

$$L_k = N_k(0) \text{ and } M_k = N_k(t_p) \tag{6.24}$$

given by

$$N_k(t) = \frac{1}{2\pi} \int_{-\infty}^{\infty} (j\omega)^k \left| H_p^*(j\omega) Z_t(j\omega) \right|^2 \cdot e^{j\omega t} d\omega \tag{6.25}$$

The error function arguments, Equations 6.2, 6.4 and 6.6, may now be expressed as

$$Q_r^2 = \frac{(bR)^2}{S_0} \frac{\left[\lambda_r(J_0 - K_0) - \lambda_s K_1 + \lambda_f(K_0 - J_1)\right]^2}{\left[-\lambda_s^2 L_2 + \left(\lambda_r^2 + \lambda_f^2\right)L_0 + 2\lambda_f\lambda_r M_0 - 2\lambda_r\lambda_s M_1\right]} \tag{6.26}$$

$$Q_t^2 = \frac{(bR)^2}{S_0} \frac{\left[\lambda_r K_0 + \lambda_f J_0\right]^2}{\left[-\lambda_s^2 L_2 + \left(\lambda_r^2 + \lambda_f^2\right)L_0 + 2\lambda_f\lambda_r M_0 - 2\lambda_r\lambda_s M_1\right]} \tag{6.27}$$

$$Q_s^2 = \frac{(bR)^2}{S_0} \left[\frac{mT_n}{2n}\right]^2 \cdot \frac{\left[-\lambda_r K_1 - \lambda_s J_2\right]^2}{\left[-\lambda_s^2 L_2 + \left(\lambda_r^2 + \lambda_f^2\right)L_0 + 2\lambda_f\lambda_r M_0 - 2\lambda_r\lambda_s M_1\right]} \tag{6.28}$$

from which the error probabilities P_s, P_r and P_f may be determined.

It is more revealing to calculate the receiver sensitivity in terms of the system variables v and t_p, where $v = (v_d / v_p)$, which are related to the Lagrangian multipliers through the following equations:

$$v = \frac{v_d}{v_p} = \frac{\lambda_r K_0 + \lambda_f J_0}{\lambda_r J_0 - \lambda_s K_1 + \lambda_f K_0} \tag{6.29}$$

and

$$v'_p = \left(\frac{d\langle v_o(t)\rangle}{dt}\right)_{t=t_p} = -\lambda_s K_2 + \lambda_f K_1 = 0 \tag{6.30}$$

Equations 6.13, 6.29 and 6.30 may be solved simultaneously and the solution written as

$$\lambda_r = \frac{\Delta_r}{\Delta}, \ \lambda_s = \frac{\Delta_s}{\Delta}, \ \lambda_f = \frac{\Delta_f}{\Delta} \tag{6.31}$$

$$\Delta = vK_1^2 + K_2(J_0 - vK_0) - (K_1 + K_2)(K_0 - vJ_0) \tag{6.32}$$

$$\Delta_r = vK_1^2 + K_2(J_0 - vK_0) \tag{6.33}$$

$$\Delta_s = K_1(K_0 - vJ_0) \tag{6.34}$$

$$\Delta_f = -K_2(K_0 - vJ_0) \tag{6.35}$$

The receiver sensitivity can now be optimised in terms of the system variables v and t_p. For a given pulse shape and assumed values of v and t_p the J and K integrals are calculated, from which the Lagrangian multipliers may be determined and hence the error probabilities in terms of the pulse energy, b. An inner iterative loop determines the value of b that satisfies the performance criterion given by Equation 6.7. The system parameters v and t_p are optimised by standard numerical techniques.

Calculations of receiver sensitivity have been performed for a system at a 1.3 μm wavelength and a bit-rate of 140 Mbit/s, using the following practical receiver parameters for a CC-CE PIN-BJT pre-amplifier [24]: receiver bandwidth of $f_c = 480$ MHz, a white noise PSD of 3.2 pA/$\sqrt{Hz}$ and transimpedance of $Z_t = 5$ kΩ. The noise PSD is assumed white, the receiver transfer function is taken to be single pole and the received pulse shape assumed Gaussian with Fourier transform pair

$$h_p(t) = \frac{1}{\sqrt{2\pi}\alpha} exp\left(\frac{-t^2}{2\alpha^2}\right) \tag{6.36}$$

$$H_p(\omega) = exp\left(\frac{-(\alpha\omega)^2}{2}\right) \tag{6.37}$$

The pulse variance α is normalised to the PCM bit period T_b [25].

The $I_k(t)$ and $N_k(t)$ integrals become

$$I_0(t) = \frac{\omega_c}{2} e^{(\alpha\omega_c)^2} e^{-\omega_c t} \text{erfc}\left(\alpha\omega_c - \frac{t}{2\alpha}\right) \tag{6.38}$$

$$I_1(t) = \frac{\omega_c^2}{2} e^{(\alpha\omega_c)^2} e^{-\omega_c t} \left[\frac{e^{-\left(\alpha\omega_c - \frac{t}{2\alpha}\right)^2}}{\alpha\omega_c \sqrt{\pi}} - \text{erfc}\left(\alpha\omega_c - \frac{t}{2\alpha}\right) \right] \tag{6.39}$$

$$I_2(t) = \frac{\omega_c^3}{2} e^{(\alpha\omega_c)^2} e^{-\omega_c t} \left[\text{erfc}\left(\alpha\omega_c - \frac{t}{2\alpha}\right) - \frac{e^{-\left(\alpha\omega_c - \frac{t}{2\alpha}\right)^2}}{(\alpha\omega_c)^2 \sqrt{\pi}} \left(\alpha\omega_c + \frac{t}{2\alpha}\right) \right] \tag{6.40}$$

$$N_0(t) = \frac{\omega_c}{4} \cdot e^{(\alpha\omega_c)^2} \cdot \left[2\cosh(\omega_c t) - e^{-\omega_c t} \cdot \text{erf}\left(\alpha\omega_c - \frac{t}{2\alpha}\right) \right.$$
$$\left. - e^{\omega_c t} \cdot \text{erf}\left(\alpha\omega_c + \frac{t}{2\alpha}\right) \right] \tag{6.41}$$

$$N_1(t) = \frac{\omega_c^2}{4} \cdot e^{(\alpha\omega_c)^2} \cdot \left[2\sinh(\omega_c t) + e^{-\omega_c t} \cdot \text{erf}\left(\alpha\omega_c - \frac{t}{2\alpha}\right) \right.$$
$$\left. - e^{\omega_c t} \cdot \text{erf}\left(\alpha\omega_c + \frac{t}{2\alpha}\right) \right] \tag{6.42}$$

$$N_2(t) = \frac{\omega_c^3}{4} \cdot e^{(\alpha\omega_c)^2} \cdot \left[e^{-\omega_c t} \left(\text{erfc}\left(\alpha\omega_c - \frac{t}{2\alpha} \right) - \frac{1}{\alpha\omega_c \sqrt{\pi}} e^{-\left(\alpha\omega_c - \frac{t}{2\alpha} \right)^2} \right) \right.$$

$$\left. + e^{\omega_c t} \left(\text{erfc}\left(\alpha\omega_c + \frac{t}{2\alpha} \right) - \frac{1}{\alpha\omega_c \sqrt{\pi}} e^{-\left(\alpha\omega_c + \frac{t}{2\alpha} \right)^2} \right) \right]$$

$$(6.43)$$

In order to calculate the probability of a false alarm error (Equation 6.5) the time at which the autocorrelation function of the noise has become small, τ_r, is required. The autocorrelation function of the noise at the output of the optimal filter is given by

$$R(\tau) \propto -\lambda_s^2 N_2(\tau) + \left(\lambda_r^2 + \lambda_f^2 \right) N_0(\tau) + 2\lambda_f \lambda_r N_0\left(\tau - t_p \right)$$

$$+ 2\lambda_r \lambda_s N_I\left(t - t_p \right)$$

$$(6.44)$$

As the autocorrelation function is dependent on the Lagrangian multipliers, τ_r will be a function of the system operating conditions (number of PPM slots, modulation index and received pulse variance). If τ_r is selected such that the normalised autocorrelation function $r(\tau) = \dfrac{R(\tau)}{R(0)}$ is small, then $\left(\dfrac{mT_n}{n\tau_r} \right)$ independent samples of the noise can be taken and the probability of false alarms determined from Equation 6.5. As a conservative estimate for calculating P_f, $\tau_r = \alpha$ will be used throughout this chapter.

6.5.2 The matched filter

In the experimental systems [15,16], the delay arm of the PDD network was realised using coaxial delay lines. As the required delay is a function of the number of PPM slots and the available channel bandwidth, this meant that the coaxial cable had to be cut to specific lengths such that results could be obtained for the various operating conditions. Additionally, the weightings of the three branch pulse shaping network are dependent upon the available bandwidth and number of PPM slots and so were varied using switched attenuators. If this section of the optimum filter can be removed then the receiver complexity would be greatly simplified. In this case, the pre-detection filter output voltage, its derivative and the mean square noise are given by

$$v_o(t) = \frac{bR}{2\pi} \int\limits_{-\infty}^{\infty} |H_p(j\omega)|^2 Z_t(j\omega) \cdot e^{j\omega t} d\omega = \frac{bR}{2\pi} \int\limits_{-\infty}^{\infty} \frac{e^{-(\alpha\omega)^2}}{(1+j\omega/\omega_c)} \cdot e^{j\omega t} d\omega$$

$$= \frac{bR\omega_c}{2} e^{(\alpha\omega_c)^2} e^{-\omega_c t} \text{erfc}\left(\alpha\omega_c - \frac{t}{2\alpha}\right) \tag{6.45}$$

$$\frac{dv_o(t)}{dt} = \frac{bR}{2\pi} \int\limits_{-\infty}^{\infty} j\omega |H_p(j\omega)|^2 Z_t(j\omega) \cdot e^{j\omega t} d\omega$$

$$= \frac{bR\omega_c^2}{2} e^{(\alpha\omega_c)^2} e^{-\omega_c t} \left[\frac{e^{-\left(\alpha\omega_c - \frac{t}{2\alpha}\right)^2}}{\alpha\omega_c \sqrt{\pi}} - \text{erfc}\left(\alpha\omega_c - \frac{t}{2\alpha}\right) \right] \tag{6.46}$$

$$\langle n_o(t)^2 \rangle = \frac{S_0}{2\pi} \int\limits_{-\infty}^{\infty} |H_p^*(j\omega)Z_t(j\omega)|^2 d\omega = \frac{S_0\omega_c}{2} e^{(\alpha\omega_c)^2} \text{erfc}(\alpha\omega_c) \tag{6.47}$$

from which the error function arguments and error probabilities can be determined.

6.5.3 Optimised 3-pole filter

In the experimental systems, the matched filter was a complex 10th order ladder structure. Here, the filter complexity is reduced by replacing the matched filter/PDD network combination with a simple 3-pole filter with pole locations $p_1 = -a_1$, $p_2 = -a_2 + jb_2$ and $p_3 = -a_2 - jb_2$ and optimising the pole locations in order to give maximum sensitivity. The output of this pre-detection filter, its derivative and the mean square noise are given by

$$v_o(t) = \frac{bR\omega_c a_1\left(a_2^2 + b_2^2\right)}{2(a_1 - \omega_c)\left[(a_2 - \omega_c)^2 + b_2^2\right]} e^{-\omega_c t} e^{\frac{(\alpha\omega_c)^2}{2}} erfc\left(\frac{\alpha\omega_c}{\sqrt{2}} - \frac{t}{\sqrt{2}\alpha}\right)$$

$$+ \frac{bR\omega_c a_1\left(a_2^2 + b_2^2\right)}{2(\omega_c - a_1)\left[(a_2 - a_1)^2 + b_2^2\right]} e^{-a_1 t} e^{\frac{(\alpha a_1)^2}{2}} erfc\left(\frac{\alpha a_1}{\sqrt{2}} - \frac{t}{\sqrt{2}\alpha}\right)$$

$$+ \frac{bR\omega_c a_1\left(a_2^2 + b_2^2\right)|z_2|}{2\left[(a_2 - \omega_c)^2 + b_2^2\right]\left[(a_2 - a_1)^2 + b_2^2\right]}\left[2e^{-a_2 t} e^{\frac{\alpha^2}{2}\left(b_2^2 - a_2^2\right)} sin[b_2 t + arg(z_2)]\right.$$

$$\left. - \alpha^2 a_2 b_2 - e^{-\frac{t^2}{2\alpha^2}}|w(z_1)| sin\left[arg(w(z_1)) + arg(z_2)\right]\right]$$

(6.47)

$$\frac{dv_o(t)}{dt} = \frac{bR\omega_c^2 a_1\left(a_2^2 + b_2^2\right)}{2(\omega_c - a_1)\left[(a_2 - \omega_c)^2 + b_2^2\right]} e^{-\omega_c t} e^{\frac{(\alpha\omega_c)^2}{2}} erfc\left(\frac{\alpha\omega_c}{\sqrt{2}} - \frac{t}{\sqrt{2}\alpha}\right)$$

$$- \frac{bR\omega_c a_1^2\left(a_2^2 + b_2^2\right)}{2(\omega_c - a_1)\left[(a_2 - a_1)^2 + b_2^2\right]} e^{-a_1 t} e^{\frac{(\alpha a_1)^2}{2}} erfc\left(\frac{\alpha a_1}{\sqrt{2}} - \frac{t}{\sqrt{2}\alpha}\right)$$

$$+ \frac{bR\omega_c a_1\left(a_2^2 + b_2^2\right)|z_3|}{2\left[(a_2 - \omega_c)^2 + b_2^2\right]\left[(a_2 - a_1)^2 + b_2^2\right]}\left[2e^{-a_2 t} e^{\frac{\alpha^2}{2}\left(b_2^2 - a_2^2\right)} sin[b_2 t + arg(z_3)]\right.$$

$$\left. - \alpha^2 a_2 b_2 - e^{-\frac{t^2}{2\alpha^2}}|w(z_1)| sin\left[arg(w(z_1)) + arg(z_3)\right]\right]$$

(6.48)

where

$$z_1 = \frac{-\alpha b_2}{\sqrt{2}} + j\left[\frac{t}{\sqrt{2}\alpha} - \frac{\alpha a_2}{\sqrt{2}}\right]$$

$$z_2 = \frac{a_1(a_2 - \omega_c) - a_2^2 + a_2 b_2 + b_2^2}{b_2} + j\left[a_1 - 2a_2 + \omega_c\right]$$

$$z_3 = \frac{a_1\left(a_2^2 - a_2\omega_c + b_2^2\right) + (\omega_c - a_2)\left(a_2^2 + b_2^2\right)}{b_2} + j\left[a_1\omega_c - a_2^2 - b_2^2\right]$$

$$w(z) = e^{-z^2}\left[1 + \frac{2j}{\sqrt{\pi}} \int_0^z e^{t^2} dt\right]$$

$$\left\langle n_o(t)^2 \right\rangle = \frac{S_0}{2\pi} \int_{-\infty}^{\infty} \frac{\left[a_1\left(a_2^2 + b_2^2\right)\omega_c\right]^2}{\left(a_1^2 + \omega^2\right)\left(a_2^2 + (\omega + b_2)^2\right)\left(a_2^2 + (\omega - b_2)^2\right)\left(\omega_c^2 + \omega^2\right)} \quad (6.49)$$

The time of the pulse peak is located by solving Equation 6.48 with respect to t giving t_p and hence v_p. The optimisation variable is taken as t_d from which v_d and v_d' can be evaluated. Knowing v_p, v_d and v_d', along with the mean-square noise, allows the error function arguments to be calculated along with the error probabilities. A standard algorithm is used to maximise the sensitivity with respect to the pole locations $p_1 = -a_1$, $p_2 = -a_2 + jb_2$ and $p_3 = -a_2 - jb_2$ and t_d.

6.5.4 Third order Butterworth filter

In this case, the pole locations are fixed to give a classical third-order Butterworth response. The -3 dB cut-off frequency of the Butterworth filter has been set to equal that of an ideal Gaussian matched filter; that is, $\omega_b = \dfrac{\sqrt{\ln(2)}}{\alpha}$. The poles become $p_1 = -\dfrac{\sqrt{\ln(2)}}{\alpha}$, $p_2 = (-\dfrac{1}{2} + j\dfrac{\sqrt{3}}{2})\dfrac{\sqrt{\ln(2)}}{\alpha}$, $p_3 = (-\dfrac{1}{2} - j\dfrac{\sqrt{3}}{2})\dfrac{\sqrt{\ln(2)}}{\alpha}$ and so the pre-detection filter output and its derivative are given by Equations 6.47 and 6.48 where $a_1 = \dfrac{\sqrt{\ln(2)}}{\alpha}$, $a_2 = \dfrac{\sqrt{\ln(2)}}{2\alpha}$ and $b_2 = \dfrac{\sqrt{3\ln(2)}}{2\alpha}$. The mean-square noise at the output of the third-order Butterworth filter is given by

$$\left\langle n_o(t)^2 \right\rangle = \frac{S_0}{2\pi} \int_{-\infty}^{\infty} \frac{1}{\left(1 + \left[\dfrac{\omega}{\omega_b}\right]^6\right)\left(1 + \left[\dfrac{\omega}{\omega_c}\right]^2\right)} d\omega \quad (6.50)$$

Again, the root of Equation 6.48 is used to locate t_p and standard numerical optimisation techniques are used to maximise the receiver sensitivity with respect to t_d.

6.6 Performance evaluation

The variation in receiver sensitivity (dBm) as a function of the number of time slots, n, for various received pulse variances (normalised to the PCM bit period) is shown in Figure 6.4. The results show that there is a value of n which optimises the receiver sensitivity. This may be explained with reference to Figures 6.5 and 6.6.

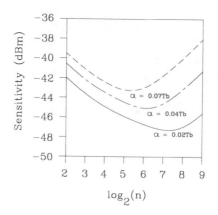

Figure 6.4 *Receiver sensitivity as a function of n when employing optimum pre-detection*

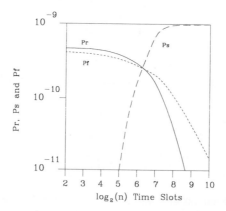

Figure 6.5 *P_r, P_s and P_f as a function of n*

These show the PPM error probabilities and the pulse shaping network components as a function of n ($\alpha=0.04T_b$ and $m=0.8$). In the low n region it can be seen from Figure 6.5 that the situation is analogous to PCM in that there is a trade-off between false alarm and erasure errors. It is well known that the optimal filter under this condition is a matched filter. However, in PCM such a filter would cause intersymbol interference. As a guard interval is left at the end of each digital PPM time frame, interframe interference is not a problem. Thus at low n a matched filter is used and the PDD network may be dispensed with as illustrated in Figure 6.6.

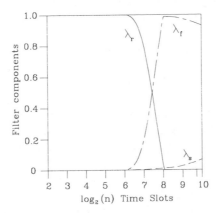

Figure 6.6 *Filter components as a function of n*

As n is increased the number of PCM bits being conveyed by the single PPM pulse also increases and consequently the number of photons per bit-time decreases. This gives rise to the improvement in receiver sensitivity displayed in Figure 6.4. However, the frame time $T_n = T_b \log_2(n)$ and so as the coding level M is increased there is more time for a false alarm error to occur (Equation 6.5). In order to counter this effect and maintain the balance between false alarm and erasure errors the threshold level is increased slightly.

The time slot duration is given by (mT_n/n) and so decreases with increasing n. Eventually a point is reached when the slot duration is comparable to the rise time of the pulse and, as Figure 6.5 shows, wrong-slot errors begin to occur. In order to curtail these errors, the rising edge of the pulse is sharpened by introducing the derivative component, reducing the delay component and increasing the proportional component of the PDD network. The performance criterion (Equation 6.7) is maintained by reducing P_f and P_r to allow for the increase in P_s. This can only be achieved by increasing the pulse energy and so the sensitivity degrades as n is further increased. Although the derivative component sharpens the pulse rise time, in order to minimise wrong-slot errors the threshold must be set at the steepest gradient of the rising edge. As such, with an increasing derivative contribution the detection threshold decreases. The net effect of this is to cause P_r to reduce at a faster rate than P_f. The combination of an increased noise equivalent bandwidth due to the derivative component (Equation 6.23) and the low optimum threshold results in sensitivity being optimised by a trade-off between wrong-slot and false alarm errors. In the high n region wrong-slot errors predominate as shown in Figure 6.5. As n is further increased the PDD network tends towards the derivative component.

From Figure 6.5 it can be seen that there is a value of n at which the three error sources are equal. Again an analogy can be drawn with PCM in that the optimum sensitivity, as shown in Figure 6.4 for $\alpha=0.04T_b$, occurs when the contributions from the error sources are equal. Above this optimum the sensitivity is maximised by balancing false alarm errors with wrong slot errors, below the optimum by balancing erasure errors with false alarm errors. In summary, the sensitivity improves with increasing n because more bits are being conveyed by the single PPM pulse. This improvement continues until the rise time of the pulse is comparable to the time slot duration and wrong slot errors occur. For this reason the optimum sensitivity displayed in Figure 6.4 increases with decreasing pulse variance. That is, a higher value of n is achievable before the onset of wrong slot errors.

Figure 6.7 compares the performance of a digital PPM system employing a Gaussian matched filter as the pre-detection filter with the optimum results of Figure 6.6 at received pulse variances of $\alpha = 0.07T_b$ and $\alpha = 0.02T_b$. In the low n region, where sensitivity is optimised by a trade-off between erasure and false alarm errors, the optimum filter reduces to a matched filter as illustrated by the values of the PDD network components in Figure 6.6 ($\lambda_r = 1$).

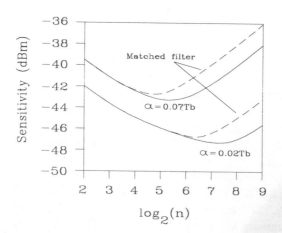

Figure 6.7 *Receiver sensitivity as a function of n with a matched filter*

Hence, as might be expected, the sensitivity calculations of section 6.6 agree with those of section 6.5. However, as *n* increases and wrong slot errors begin to occur, the only corrective action available when operating with a matched filter is to vary the detection threshold level *v* whereas, with the optimum filter, the PDD network is used to shape the pulse leading to the improved sensitivities displayed in Figure 6.7. At $\alpha = 0.02T_b$ the optimum system offers a sensitivity of -47.4 dBm when operating with 128 slots. The maximum sensitivity offered by the matched filter system is -47 dBm which is within 0.4 dB of the optimum. This illustrates that the complicated pulse shaping network may dispensed with. This has been verified experimentally by Massarella and Sibley [16] for a 16-ary PPM system with a time slot duration of 20 ns.

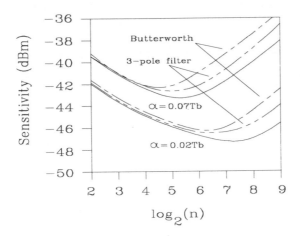

Figure 6.8 *Receiver sensitivity as a function of n with sub-optimum detection*

The sensitivity results of the optimised 3-pole and the third-order Butterworth filters are illustrated in Figure 6.8. At low values of *n*, they are within 0.1 dB and 0.35 dB of the optimum pre-detection filter and so are in good agreement. As *n* is increased, the Butterworth filter results in a more rapid degradation in sensitivity than that of the matched filter (Figure 6.7). This occurs due to the increased pulse dispersion caused by the Butterworth filter. At high values of *n*, the optimised 3-pole filter offers better sensitivity than both the Butterworth and matched filter systems. This is because the pole-locations are optimised in order to shape the pulse so as to limit the occurrence of wrong-slot errors. The disadvantage of this filter is that the pole-locations, and hence component values, are dependent upon *n*. Figure 6.9 indicates the cost (dB) of employing these sub-optimum filters in comparison to the optimum pre-detection filter. At low values of *n*, all three filters offer sensitivities that are within 0.35 dB of the optimum pre-detection filter. As wrong slot errors begin to occur, both the Butterworth and the matched filter result in a fairly rapid degradation in sensitivity. At high values of n, the optimised 3-pole filter results in the least degradation. This is because there is flexibility in placing the poles in order to balance false alarm and wrong slot errors. Figure 6.9 shows that at very high values of n the degradation of each sub-optimum filter is tending to a constant value. This results because the pulse shaping optimum filter is tending towards the derivative component and so no further improvement in the sensitivity is achievable.

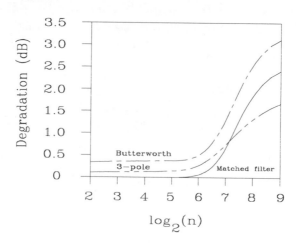

Figure 6.9 *Degradation in receiver sensitivity due to sub-optimum detection*

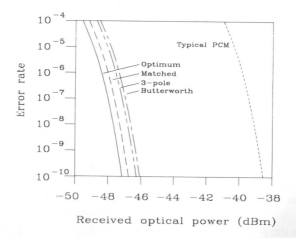

Figure 6.10 *Error rate curves for $\alpha=0.02T_b$*

Figure 6.10 shows the receiver error rate plotted as a function of the received optical power (dBm) for all the pre-detection filter systems considered in this chapter along with the error rate curve for a typical 140 Mbit/s, 1.3 μm PCM system [26]. The maximum sensitivities of digital PPM employing the optimal filter, matched filter, optimised 3-pole filter and the third-order Butterworth filter are -47.4 dBm, -47 dBm, -46.5 dBm and -46.3 dBm respectively. The sub-optimum filters offer sensitivities that are within 0.4 dB, 0.9 dB and 1.1 dB of that offered by the optimum pre-detection filter. The typical PCM system offers a sensitivity of -38.8 dBm and so digital PPM gives an improvement of 8.6 dB. This reduces to 7.5 dB when operating with a third-order order Butterworth filter. These results clearly illustrate that receiver complexity can be significantly simplified without enormous degradations in receiver sensitivity. Further, a simple Butterworth pre-detection filter can be used with only a 1.1 dB reduction in sensitivity. The performance improvement achieved by utilising PPM can be significantly reduced if timing impairments prevail. This is attributed to the temporal nature of the PPM format in that the information is transmitted as the discrete time position of the PPM pulse in the time frame.

The remainder of this chapter is devoted to the investigation of the synchronisation requirements and impairments in PPM. In particular, slot and frame synchronisation will be analysed and the performance impairment associated with non-ideal self-synchronisation will be assessed. The optimum set of parameters that reduces this impairment will be establised.

6.7 Spectral characterisation and slot synchronisation

In this section a spectral characterisation is presented for the cyclostationary optical fibre digital PPM format. The characterisation is utilised to study slot synchronisation in PPM. A linearised phase lock loop model (PLL) model is then employed in conjunction with the characterisation to evaluate the impairment due to self slot synchronisation.

The digital PPM structure is shown in Figure 6.11 together with some naturally occurring sequences (to be discussed in section 6.8). The frame and slot duration are T_f and $(mT_f)/n$ respectively. Mathematically the digital PPM pulse train can be represented by

$$x(t) = \sum_{k=-\infty}^{\infty} g(t - kT_f - t_k) \tag{6.51}$$

where $g(t)$ is the pulse shape; t_k is a stochastic sequence defining the pulse positions and is stationary in the wide sense. The digital PPM pulse train in Equation 6.51 can be viewed as a stationary process subjected to a repetitive periodic processing operation. This can be treated as a cyclostationary process [27,28]. According to the Wiener-Kintchine theorem, the power spectral density

(PSD) of such a process can be evaluated by first determining the autocorrelation function. Rather than the classic time averaging, statistics of cyclostationary processes can be derived by forming a related stationary process from which time-invariant statistics can be obtained directly [29]. The related stationary process can be obtained by randomising the time reference for the individual realisations of Equation 6.51, then averaging the required statistics over the ensemble formed by the set of realisations.

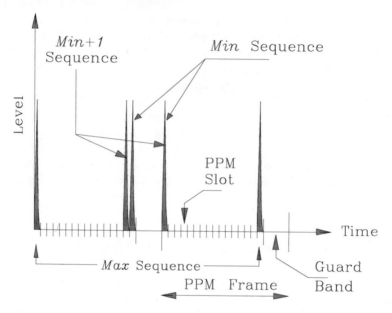

Figure 6.11 *The PPM signal structure and the sequences tracked*

Introducing the phase-randomising variable θ uniformly distributed in $\{0, T_f\}$, the equivalent stationary process will be

$$x(t) = \sum_{k=-\infty}^{\infty} g(t - kT_f - t_k - \theta) \qquad (6.52)$$

The autocorrelation, the statistic of interest, can then be evaluated by averaging over the ensemble in the period $\{0, T_f\}$. Further, due to the time-invariant properties the general correlation $\mathcal{R}(\tau)$ can be expressed as

$$\mathcal{R}(\tau) = \frac{1}{T_f} \sum_{k=-\infty}^{\infty} R_k(\tau - kT_f) \tag{6.53}$$

where $R_k(\tau)$ is the correlation between two frames, k frames apart, and is given by

$$R_k(\tau) = \sum_{j=0}^{N-1} R_{kj}\left\{\tau - j\frac{T_f}{N}\right\} \tag{6.54}$$

$N = n/m$ being the total length of the frame in slots. Equation 6.53 represents frame decomposition while Equation 6.54 represents slot decomposition. The partial autocorrelation $R_k(\tau)$ is given by

$$R_k(\tau) = < g(t_0 + \theta)g(t_k + \tau + \theta) >_{t_o, t_k, \theta} \tag{6.55}$$

where θ is random and uniform in $\{0, T_f\}$, t_0 represents the position of the pulse in frame 0 and t_k the pulse position in frame k. Making the ensemble averages explicit gives

$$R_k(\tau) = \frac{1}{T_f} \int_0^{T_f} d\theta \int_0^{T_f} dt_0 \int_0^{T_f} dt_k \int_{-\infty}^{\infty} dt\, p(t_o, t_k) g(t - t_0 + \theta) g(t - t_k + \tau + \theta) \tag{6.56}$$

where $p(t_o, t_k)$ is the joint probability. Assuming that t_0 and t_k are a set of independent identically distributed random variables then

$$p(t_0, t_k) = p(t_0)p(t_k) \tag{6.57}$$

For digital PPM the probability is discrete by virtue of the allowed pulse positions, hence

$$p(t_0, t_k) = \sum_{i=0}^{N-1} a_i \delta\left(t_0 - \frac{iT_f}{N}\right) \sum_{l=0}^{N-1} a_l \delta\left(t_k - \frac{lT_f}{N}\right) \tag{6.58}$$

where a_i and a_l are array elements that represent the probability distribution in the frames. The pulse shape $g(t)$ can be represented by an indicator function:

$$g(t) = X_{\left(-\frac{\Delta}{2}, \frac{\Delta}{2}\right)}(t) \tag{6.59}$$

where

$$X_{(I)}(t) = \begin{cases} 1 & \text{if } t \in I \\ 0 & \text{otherwise.} \end{cases} \qquad (6.60)$$

The correlation integral becomes

$$\frac{1}{T_f} \int_{-\infty}^{\infty} dt\, g(t - t_0 + \theta) g(t - t_k + \tau + \theta) =$$

$$\frac{1}{T_f} \int_{-\infty}^{\infty} dt\, X_{\left(t_0-\theta-\frac{\Delta}{2}, t_0-\theta+\frac{\Delta}{2}\right)}(t) X_{\left(t_k-\tau-\theta-\frac{\Delta}{2}, t_k-\tau-\theta+\frac{\Delta}{2}\right)}(t) \qquad (6.61)$$

$$= \frac{1}{T_f} \int_{-\infty}^{\infty} dt\, X_{\left(t_0-\theta-\frac{\Delta}{2}, t_0-\theta+\frac{\Delta}{2}\right) \cap \left(t_k-\tau-\theta-\frac{\Delta}{2}, t_k-\tau-\theta+\frac{\Delta}{2}\right)}(t)$$

$$= \begin{cases} \dfrac{1}{T_f}\left[\Delta - \left|\tau - (t_k - t_0)\right|\right] & \text{if } \left|\tau - (t_k - t_0)\right| < \Delta \\ 0 & \text{otherwise.} \end{cases} \qquad (6.62)$$

θ vanishes and the corresponding integral is reduced. Inserting the expressions for $p(t_o, t_k)$, carrying out the integrations and making the appropriate δ-function substitutions gives

$$R_k(\tau) = \frac{1}{T_f} \sum_{i=0}^{N-1} \sum_{l=0}^{N-1} a_i a_l q(l - i, \tau) \qquad (6.63)$$

where $q(j, \tau)$ is

$$q(j, \tau) = \begin{cases} \left[\Delta - \left|\tau - j\dfrac{T_f}{N}\right|\right] & \text{if } \left|\tau - j\dfrac{T_f}{N}\right| < \Delta \\ 0 & \text{otherwise.} \end{cases} \qquad (6.64)$$

Carrying out the summations in terms of j gives

$$R_k(\tau) = \frac{1}{T_f} \sum_{j=-N+1}^{N-1} \sum_{l=0}^{N-1-|j|} a_l a_{l+|j|} \rho\left(\tau - j\frac{T_f}{N}\right) \tag{6.65}$$

where

$$\rho(u) = \begin{cases} [\Delta - |u|] & \text{if } |u| < \Delta \\ 0 & \text{otherwise.} \end{cases} \tag{6.66}$$

Evaluating $R_0(\tau)$ separately, in a similar manner

$$R_0(\tau) = \frac{1}{T_f} \int_0^{T_f} d\theta \int_0^{T_f} dt_0 \int_{-\infty}^{\infty} dt\, p(t_o) g(t - t_0 + \theta) g(t - t_0 + \tau + \theta) \tag{6.67}$$

which gives

$$R_0(\tau) = \sum_{k=0}^{N-1} a_k \rho(\tau) \tag{6.68}$$

The autocorrelation for the digital PPM pulse stream is then

$$\mathcal{R}(\tau) = \frac{1}{T_f}\left\{ R_0(\tau) + \sum_{\substack{k=-\infty \\ k\neq0}}^{\infty} R_k\left(\tau - kT_f\right) \right\} \tag{6.69}$$

which becomes

$$\mathcal{R}(\tau) = \frac{1}{T_f}\left\{ \rho(\tau) + \sum_{\substack{k=-\infty \\ k\neq0}}^{\infty}\left(\sum_{j=-N+1}^{N-1} \sum_{l=0}^{N-1-|j|} a_l a_{l+|j|} \rho\left(\tau - j\frac{T_f}{N} - kT_f\right) \right) \right\} \tag{6.70}$$

Using cyclic correlation reduces $\mathcal{R}(\tau)$ to

$$R(\tau) = \frac{1}{T_f} \left\{ \rho(\tau) + \sum_{\substack{k=-\infty \\ k \neq 0}}^{\infty} \left(\sum_{j=0}^{N-1} \beta_j \, \rho\left(\tau - j\frac{T_f}{N} - kT_f\right) \right) \right\} \tag{6.71}$$

and adding and subtracting the term $k=0$ gives

$$R(\tau) = \frac{1}{T_f} \left\{ \rho(\tau) - \sum_{j=0}^{N-1} \beta_j \, \rho\left(\tau - j\frac{T_f}{N}\right) + \sum_{k=-\infty}^{\infty} \sum_{j=0}^{N-1} \beta_j \, \rho\left(\tau - j\frac{T_f}{N} - kT_f\right) \right\} \tag{6.72}$$

where β is the cyclic correlation $\bar{\beta} = \bar{a} \otimes \bar{a}$. Taking the Fourier transform of this gives an original expression for the PSD of the optical fibre digital PPM pulse stream:

$$S(f) = |G(f)|^2 \left\{ \frac{1 - |B(f)|^2}{T_f} \right\} + \frac{1}{T_f^2} \left\{ \sum_{k=-\infty}^{\infty} \left| G\left(\frac{k}{T_f}\right) \right|^2 B_s \delta\left(f - \frac{k}{T_f}\right) \right\} \tag{6.73}$$

where

$G(f)$ is the Fourier transform of the pulse shape $g(t)$
B is the modulus squared of the characteristic function of the data
distribution and s is a selector given by $s = \left(k + \dfrac{N}{2}\right)\mathrm{mod}(N) - \dfrac{N}{2}$

The characteristic function $B(f)$ due to the PPM probability distribution on the frame (including allowed and guard band) is given by

$$B(f) = \sum_{i=0}^{N-1} a_i e^{j2\pi f \frac{T_f}{N}} \tag{6.74}$$

The first term of Equation 6.73 represents the continuum associated with the random PPM signalling while the second term represents discrete components at the PPM frame repetition rate (and hence slot rate). The slot synchroniser consists of a pulse shaping network (where the input pulses are processed into half slot duration pulses) followed by a phase lock loop (with bandwidth B_L) that tracks the discrete slot rate component. Using a linearised PLL model, the timing variance of the output extracted clock is given in rad^2 by [30]

$$\sigma_s^2 = \frac{B_L \mathcal{N}}{P_s} \tag{6.75}$$

where $\mathcal{N}$ is the noise power (the continuum in the fundamental case) and P_s is the discrete slot rate power. In general, since the noise is not white in the vicinity of the slot discrete component, the slot timing variance can be written as

$$\sigma_s^2 = \int_{F_s - \frac{B_L}{2}}^{F_s + \frac{B_L}{2}} \left\{ 1 - \left| \sum_{i=0}^{N-1} a_i e^{j2\pi f \frac{T_f}{N}} \right|^2 \right\} df \tag{6.76}$$

The B_s term of Equation 6.73 is unity at the slot rate. The jittered slot clock can cause the PPM pulse to be interpreted as being in the slot directly before or after the proper slot containing the pulse. This gives rise to PPM jitter wrong slot errors (JWSE). The probability of such an event, to a first approximation, is given by

$$P_{Js} = \text{erfc}\left\{ \frac{Q_{Js}}{\sqrt{2}} \right\} \tag{6.77}$$

where $\text{erfc}\{\cdot\}$ is the complementary error function and

$$Q_{Js}^2 = \left\{ \frac{\pi}{\sigma_s} \right\}^2 \tag{6.78}$$

The extracted clock timing variance was evaluated at various PPM orders (values of n) and PLL bandwidths at a modulation depth $m=0.8$. The result is shown in Figure 6.12. This clearly indicates that reduction in the PLL bandwidth would result in improved timing variance.

In the practical system, the PLL bandwidth would also be selected dependent on the required dynamic performance [30]. The desired timing variance at the PLL output is also a function of the PPM order, n. This is clearly illustrated in Figure 6.12. Lower PPM orders (e.g. $n=4$) result in a better timing variance for a given PLL bandwidth. A physical explanation can be offered to the poor performance at higher n values for a given PLL bandwidth, in that the higher the value of n is, the longer will be the expected average run of zeros at the PLL input. Therefore for a given timing performance (timing variance) Figure 6.12 offers the combinations of PPM order and PLL bandwidth.

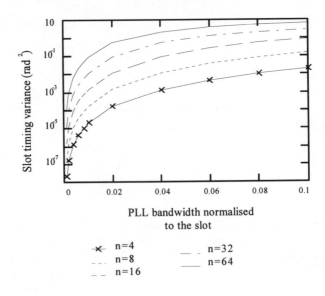

Figure 6.12 *The extracted slot clock timing variance at m=0.8*

The probability of the PPM slot clock jitter causing wrong slot errors was evaluated as a function of the PLL bandwidth for various values of the PPM order *n*. The result is shown in Figure 6.13.

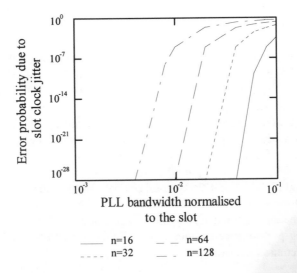

Figure 6.13 *Probability of wrong slot errors due to the slot clock jitter*

Higher PPM orders have a worse error performance and require narrower PLL bandwidth to maintain performance. This can be seen from the fact that, for a given PLL bandwidth, as n increases the extracted clock timing variance increases (Figure 6.12) and hence the probability of wrong slot error. Generally the optical fibre PPM system performance (sensitivity etc) is evaluated at an error probability of 10^{-9}. For a given required error probability (e.g. if it is required that $P_{Js} = 10^{-9}$) and a given PPM order, Figure 6.13 indicates the proper PLL bandwidth that should be used to maintain the performance.

6.8 Frame synchronisation

The PPM PSD as in Equation 6.73 contains discrete components at the frame rate; however, the variation in pulse position from one frame to the next causes large disparities and the frame phase information cannot be extracted directly from the pulse train. Neither can it be extracted by processing the PSD in Equation 6.73 since the latter is phase blind. Line coding techniques have been suggested as a method to limit the pulse disparity [31], however these methods lead to added redundancy and complexity in the form of coders and decoders.

We propose tracking naturally occurring PPM sequences to extract the frame phase information. Two distinct naturally occurring sequences suitable for frame phase extraction are shown in Figure 6.11. The minimum length sequence (*Min*) occurs when there is a PPM pulse in the last slot in one frame followed by a pulse in the first slot of the next frame. This is a distinct and phase-bearing event since it represents minimum PPM to PPM pulse separation. Another possible sequence is the maximum length sequence (*Max*). This is also a distinct and a phase-bearing event since it represents maximum PPM to PPM pulse separation. A synchroniser that tracks the class of minimum length sequences is illustrated in Figure 6.14.

A shift register with cells equal to the guard band duration plus two cells tracks the minimum length sequence. The AND gate identifies the sequence and places a pulse in the second PPM slot in the frame. However since the tracked sequences are naturally occurring events, the probability of identifying the frame phase deteriorates fast with increase in the PPM order (the joint probability for the event assuming uniform independent identically distributed (iid) PPM pulse positions is $1/n^2$).Therefore additional sequences are tracked to improve the performance. Figure 6.14 shows the topology used to track *Min*+1 and *Min*+2 sequences. A *Min*+i sequence occurs when there is a pulse at slot $(n-i-1)$ in one frame and a pulse at the first slot of the next frame. Since the *Min*+i are not distinct events (e.g. there are two *Min*+1 sequences), the required *Min*+i sequence that is in phase with the *Min* sequence is synchronously selected by the mod $\{n/m\}$ counter and the combinational logic topology of Figure 6.14.

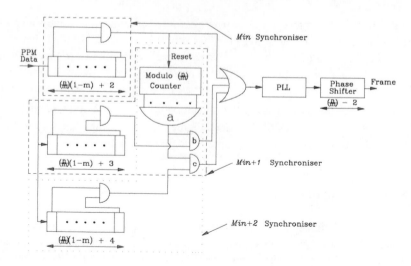

Figure 6.14 *The {Min, Min+1 and Min+2} sequences frame synchroniser*

The synchroniser is followed by a PLL that extracts the frame frequency. A static phase shift of $\{(n/m)-2\}$ slots is carried out following the PLL to restore the frame phase. The measured PSD at the PLL input due to the *Min* synchroniser alone, then due to the combined [*Min, Min*+1 and *Min*+2] synchroniser is shown in Figures 6.15 and 6.16 respectively. Figure 6.16 shows that tracking two additional sequences results in a measured 10.8 dB improvement in the frame discrete component.

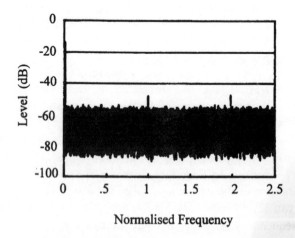

Figure 6.15 *The measured Min synchroniser PSD*

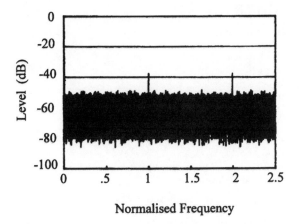

Figure 6.16 *The measured {Min, Min+1 and Min+2} synchroniser PSD*

The PSD at the output of the synchroniser can be evaluated by properly modelling the action of the synchroniser's. The synchroniser's output pulse train can be written as

$$y(t) = \sum_{k=-\infty}^{\infty} g(t - t_k) \tag{6.79}$$

where $g(t)$ is the synchroniser output pulse shape and t_k is a random variable that identifies the occurrence of the required sequence(s) and is such that

$$t_k = n_k' T_f + \frac{T_f}{N} \tag{6.80}$$

in which T_f is the PPM time frame duration, $N = n/m$ is the total number of slots in the frame and n_k' is a random number that represents the kth frame at which the required sequence was identified. The synchroniser places a pulse at the second PPM slot. The mathematical model developed here considers the synchroniser output spectrum as composed of the sum of contributions from a set of delayed pulses. The pulse delays are random and follow a rule dictated by the probability distribution of the incoming data and the sequences tracked. For a given vector t_k, $y(t)$ in Equation 6.79 represents a realisation of the synchroniser output process. If the process is to be analysed practically it is instructive to evaluate the power spectral density of a truncated realisation. We define a truncated realisation $y_T(t)$ of the synchroniser output process as:

$$y_T(t) = \begin{cases} y(t) & \text{if} \quad 0 \le t \le T \\ 0 & \text{otherwise} \end{cases} \tag{6.81}$$

such that

$$y(t) = \lim_{T \to \infty} y_T(t) \tag{6.82}$$

This truncation enables us to evaluate the Fourier transform

$$Y_T(f) = \int_{-\infty}^{\infty} y_T(t)e^{-j2\pi ft} dt \tag{6.83}$$

with $y(t)$ as a power signal, the power spectrum is associated with an area equal to the average power. Hence it is appropriate to consider the power in $y(t)$ in $\{0, T\}$:

$$P_T = \frac{E_T}{T} = \int_{-\infty}^{\infty} \frac{1}{T}|Y_T(f)|^2 df \tag{6.84}$$

The function under the integral in Equation 6.84 describes the distribution of the power with frequency. This is nothing but the required power spectrum of the truncated realisation of $y(t)$. The power spectrum for the realisation is obtained by taking the limit as below

$$S_y(f) = \lim_{T \to \infty} \frac{1}{T}|Y_T(f)|^2 \tag{6.85}$$

Since Equation 6.85 describes the power spectrum of an individual realisation of the synchroniser output process, it constitutes a random variable with respect to the process. To obtain a characterisation for the whole process one has to average over all the realisations performing the limiting operation last [32].

$$S_{Ey}(f) = \lim_{T \to \infty} \frac{1}{T} E\left\{|Y_T(f)|^2\right\} \tag{6.86}$$

Although, strictly, Equation 6.86 represents the required characterisation of the process (and describes the steps to be followed after Equation 6.79), the result of Equation 6.85 will be used with some assumptions. Firstly the limit as T approaches infinity will be taken as that when T is much greater than $n^2 T_f$ $\left(T \gg n^2 T_f\right)$; this lends itself to numeric implementation. Secondly the results of the spectral characterisation of a realisation as in Equation 6.85 will be compared directly with our measurements without averaging or carrying out the expectations. The deviation in so doing will be shown to be negligibly small once the theoretical and practical results are compared. Following these steps the spectral characterisation of a realisation of the synchroniser output signal process will be

$$S_y(f) = \frac{1}{LT_f}\left\{\left\|G(f)\sum_{k=0}^{L}e^{-j2\pi fT_f\left(n_k'+\frac{1}{N}\right)}\right\|^2\right\} \tag{6.87}$$

where $G(f)$ is the synchroniser output pulse shape transform, n_k' is a random number that represents the kth frame at which the required sequence was identified. L gives the length of the truncated sequence and is such that $L >> n^2$.

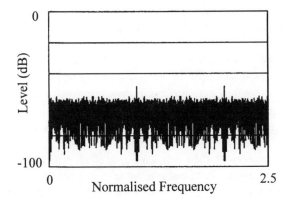

Figure 6.17 *Predicted Min synchroniser PSD*

A set of uniform random pulse positions n_k was used to represent the PPM data stream. The required sequences were tracked by computer. The random variable n_k' that represents the *Min* synchroniser output was set to be non-zero if and only if two consecutive values of n_k were $(n-1)$ and 0. Similarly the variable n_k'' that represents the [*Min*, *Min*+1 and *Min*+2] synchroniser output was set to be non-zero if and only if two consecutive values of n_k were [$(n-1)$ and 0], [$(n-2)$ and 0] or [$(n-3)$ and 0]. The variables n_k' and n_k'' were used in Equation 6.87 to yield the PSD the output of the *Min* and [*Min*, *Min*+1 and *Min*+2] synchronisers.

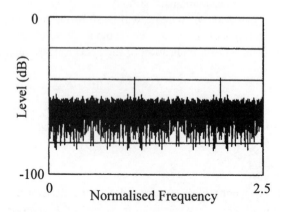

Figure 6.18 *Predicted {Min, Min+1 and Min+2} synchroniser PSD*

The results are shown in Figures 6.17 and 6.18 for the *Min* and [*Min, Min+1* and *Min+2*] synchronisers respectively. Discrete components at the frame rate and the continuum due to data randomness are clear. Figure 6.18 shows 10.2 dB improvement in the discrete frame rate component over Figure 6.17. Further, comparison of Figure 6.15 with Figure 6.17, and Figure 6.16 with Figure 6.18, shows close agreement between the theoretical and practical results. A linearised PLL model can be used to evaluate the frame timing variance σ_f^2 as in Equation 6.75 where the noise (continuum) and the discrete components can be obtained from Equation 6.37 and are depicted in Figures 6.17 and 6.18 for the two synchronisers.

The jitter associated with the extracted frame clock can cause wrong slot errors if it is such as to cause the PPM pulse to be interpreted in the slot directly before or after the proper slot. The probability of such an event can be approximated by

$$P_{Jf} = \text{erfc}\left\{\frac{Q_{Jf}}{\sqrt{2}}\right\} \qquad\qquad (6.88)$$

where

$$Q_{Jf}^2 = \left\{\frac{\pi}{\sigma_f}\right\}^2 \qquad\qquad (6.89)$$

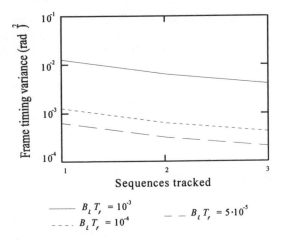

Figure 6.19 *The extracted frame clock timing variance*

The extracted frame clock variance σ_f^2 was evaluated for various PLL bandwidths and a number of tracked sequences. The result is shown in Figure 6.19. The extracted frame clock variance can be reduced by either tracking additional sequences or by reducing the PLL bandwidth. In most practical cases the optimisation will be through the use of both parameters.

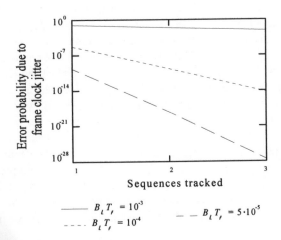

Figure 6.20 *Probability of PPM wrong slot error due to frame clock jitter*

The number of sequences tracked is selected to give the required dynamic performance (i.e. with more sequences tracked the updates to the PLL will be more frequent) and then the proper PLL bandwidth is selected for the required timing variance. The value of the timing variance would be selected in a manner that minimises the PPM JWSE due to the frame clock jitter. This probability of error is shown in Figure 6.20. It is a function of the number of tracked sequences and the PLL bandwidth. The results in Figure 6.20 would enable the selection of the optimum frame synchronisation parameters that would result in a synchronisation impairment that is within bounds and less than the system running error probability.

6.9 Summary

A performance analysis has been presented for digital PPM transmitted over an optical fibre channel and detected using both optimum and sub-optimum pre-detection filters. Receiver sensitivity calculations, carried out at a bit-rate of 140 Mbit/s and a wavelength of 1.3 μm, show that the optimum digital PPM system considered offers an 8.6 dB improvement over a typical PCM system.

The sub-optimum pre-detection filters considered were a matched filter, an optimised 3-pole filter and a third-order Butterworth filter. These led to sensitivity degradations of 0.4 dB, 0.9 dB and 1.1 dB respectively. This clearly illustrates that receiver complexity can be simplified without large reductions in sensitivity. In particular, the well known and simple Butterworth filter can be employed with only 1.1 dB degradation in sensitivity.

The timing requirements for digital optical fibre PPM have been analysed. An original spectral characterisation of the PPM format using its cyclostationary properties has been presented. The characterisation was used to evaluate the inherent systematic jitter associated with the extracted slot clock. An optimisation of the extracted slot clock timing variance and system wrong slot errors (due to imperfect slot synchronisation) was shown to be feasible in terms of the PLL bandwidth and the PPM order. Frame synchronisation was analysed for an original class of frame synchronisers that utilises natural sequences. The extracted frame clock timing variance was evaluated and the probability of wrong slot errors due to the non-ideal frame clock was assessed. The frame clock timing variance and wrong slot errors were shown to be minimisable provided that the proper number of natural sequences is tracked and the appropriate PLL bandwidth is utilised. The analysis has provided a performance evaluation of the optical fibre PPM system in the presence of inherent systematic slot and frame jitter.

6.10 References

1. Sato, M., Murata, M., and Namewaka, T.: 'A new optical communications system using the pulse interval and width modulation code', *IEEE Trans.*, 1979, **CATV-4**, pp.1-9

2. Heatley, D.: 'Video transmission in optical fibre local area networks using pulse time modulation', 9th European Conf. on Optical Communications, Geneva, 1985, pp.343-345

3. Wilson, B. and Ghassemlooy, Z.: 'Optical fibre transmission of multiplexed video signals using pulse width modulation', *Int. J. of Optoelectronics*, 1989, 4, pp.3-17

4. Wilson, B. and Ghassemlooy, Z.: 'Optical pulse width modulation for electrically isolated analogue transmission', *J. Phys E: Instrum.*, 1985, **18**, pp.954-958

5. Ghassemlooy, Z. and Wilson, B.: 'Analogue optical fibre modulation techniques for transmission of high frequency analogue signals', 8th Int. Conf. on Systems Eng., Coventry, 1991, pp.596-602

6. Garrett, I.: 'Pulse-position modulation for transmission over optical fibers with direct or heterodyne detection', *IEEE Trans.*, 1983, **COM-31**, pp.518-527

7. Garrett, I.: 'Digital-pulse position modulation over dispersive optical fibre channels', International Workshop on Digital Communications, Tirrenia, Italy, 15-19 Aug., 1983

8. Garrett, I.: 'Digital-pulse position modulation over slightly dispersive optical fibre channels', International Symposium on Information Theory, St. Jovite, pp.78-79, 1983

9. Pires, J.J.O. and Da Rocha, J.R.F.: 'Digital pulse position modulation over optical fibres with avalanche photodiode receivers', *IEE Proc. J.*, 1986, **133**, pp.309-313

10. Cryan, R.A., Unwin, R.T., Garrett, I., Sibley, M.J.N., and Calvert, N.M.: 'Optical fibre digital pulse-position-modulation assuming a Gaussian received pulse shape', *IEE Proc. J.*, 1990, **137**, pp.89-96

11. Massarella, A.J.M. and Sibley, M.J.N.: 'Optical digital pulse position modulation : Experimental results for heterodyne detection using sub-optimal filtering', *Electron. Lett.*, 1992, **28**, pp.574-575

12. Garrett, I., Calvert, M.N., Sibley, M.J.N., Unwin, R.T., and Cryan, R.A.: 'Optical fibre digital pulse position modulation', *Brit. Telecom. Technol. J.*, 1989, 7, pp.5-11

13. Cryan, R.A., Unwin, R.T., Massarella, A.J., Sibley, M.J.N., and Garrett, I.: 'Coherent Detection: *n*-ary PPM vs. PCM', SPIE Proc. on Coherent Lightwave Communications, San Jose, California, 16-21 September, 1990

14. Cryan, R.A., Unwin, R.T., Massarella, A.J., and Sibley, M.J.N.: 'Performance Analysis of a Homodyne Digital PPM System', IEE 2nd Bangor Communications Symposium, Bangor UK, 23-24 May 1990

15. Calvert, M.N., Sibley, M.J.N., and Unwin, R.T.: 'Experimental optical fibre digital pulse-position modulation system', *Electron. Lett.*, 1988, **24**, pp.129-131

16. Massarella, A.J.M., and Sibley, M.J.N.: 'Experimental results on sub-optimal filtering for optical digital pulse-position modulation', *Electron. Lett.*, 1991, **27**, pp.1953-1954

17. Elmirghani, J.M.H., Cryan, R.A., and Clayton, F.M.: 'Analytic and numerical modelling of the optical fibre PPM slot and frame spectral properties with application to timing extraction', *IEE Proc. Communications*, 1994, **141**, pp.379-389

18. Elmirghani, J.M.H., Cryan, R.A., and Clayton, F.M.: 'Spectral characterisation and frame synchronisation of optical fibre digital PPM', *Electron. Lett.*, 1992, **28**, pp.1482-1483

19. Elmirghani, J.M.H., Cryan, R.A., and Clayton, F.M.: 'PPM phase-bearing events for direct frame phase extraction', *Electron. Lett.*, 1993, **29**, pp.775-777

20. Elmirghani, J.M.H., Cryan, R.A., and Clayton, F.M.: 'Slot and frame synchronisation in optical fibre PPM systems with dispersed pulse shapes', *J. of Optical Communications*, 1994, **15**, pp.5-10

21. Elmirghani, J.M.H. and Cryan, R.A.: 'Jitter implications on PPM utilising optical preamplification', *Electron. Lett.*, 1994, **30**, pp.60-62

22. Helstrom, C.W.: 'Statistical theory of signal detection', 2nd Ed., (Pergamon Press), 1968, p.309

23. Proakis, J.J.: 'Digital Communications', (McGraw-Hill) 1983, p.152

24. Yamashita, K. *et al*: 'Simple common-collector full-monolithic preamplifier for 560 Mbit/s optical transmission', *Electron. Lett.*, 1986, **22**, pp.146-147

25. Personick, S.D.:'Receiver design for digital fiber optic communication systems, I, II', *Bell Syst. Tech. J.*, 1973, **52**, pp.843-886

26. Sibley, M.J.N., Unwin, R.T., and Smith, D.R.: 'The design of pin-bipolar transimpedance pre-amplifiers for optical receivers', *J. IERE*, 1985, **55**, pp.104-110

27. Ogura, H.: 'Spectral representation of a periodic nonstationary random process', *IEEE Trans. on Inform. Theory*, 1971, **IT-17**, pp.143-149

28. Gardner, W.A., and Franks, L.E.: 'Characterisation of cyclostationary random signal processes', *IEEE Trans. on Inform. Theory*, 1975, **IT-21**, pp.4-14

29. Hurd, H.L.: 'Stationarising properties of random shifts', *SIAM J. Appl. Math.*, 1974, **26**, pp.203-212

30. Gardner, F.M.: 'Phase lock techniques', (New York : John Wiley) 1979, Chapters 3 and 11

31. O'Reilly, J.J. and Yichao, W.: 'Line code design for digital pulse-position modulation', *IEE Proc. F*, 1985, **132**, pp. 441-446

32. Middleton, D.: 'An Introduction to Statistical Communication Theory', (McGraw-Hill) 1960, Chapter 3

Chapter 7

Performance assessment of subcarrier multiplexed optical systems

H. M. Salgado and J. J. O'Reilly

7.1 Introduction

The work presented here is concerned with an investigation of the impact of source nonlinearity on the performance of multichannel subcarrier multiplexed optical fibre systems. There has been much interest recently in the capabilities of subcarrier multiplexing (SCM) as a means of realising economical multichannel systems which can be deployed in the short to medium term to support a wide range of analogue and digital services [1, 2, 3, 4]. The attractive feature of SCM is that it provides a way of exploiting the multi-gigahertz bandwidth potential of high speed lasers [5, 6] using conventional and established microwave techniques. Also it is very flexible being capable of simultaneously transmitting conventional baseband and microwave signals with the same fibre and detector [7, 8]. Moreover, SCM systems can be combined with coherent techniques [9, 10, 11] and wavelength division multiplexing [12, 13] to utilise fully the tens of terahertz capacity of single-mode fibre.

The performance of such systems can be impaired very significantly by nonlinearities associated with the laser diode source. Realisation of the full system capabilities calls for detailed modelling and analysis of nonlinearities and associated intermodulation distortion with a view to practical system optimisation through an appropriate balance between signal-induced distortion and noise effects.

The basic distortion-producing phenomena have been studied extensively but such analytical treatments presented in the literature have concentrated on supporting experimental observations relating to relatively simple (two-tone and three-tone modulation) distortion measurements [14, 15, 16]. So far little work has been reported aimed at providing rigorous yet tractable design and analysis tools and techniques appropriate for assessing systems of the kind and complexity being envisaged for planned deployments. Indeed, some of the methods extant in the literature are not realistically extensible to cases of practical interest [17, 18]. Here we discuss and demonstrate the applicability of tools and techniques appropriate for practical systems design and optimisation.

In section 7.2 an overview of SCM systems is given and specific applications are discussed that illustrate the advantages and disavantages of the technique. The sources of noise related to the laser, intensity noise and nonlinear distortion, are then examined in section 7.3. The physical mechanisms which give rise to laser distortion and intensity noise are discussed and their impact on typical SCM systems is addressed. In this chapter we emphasise the intrinsic dynamic nonlinearity of the laser for which models are presented in section 7.4. In particular, Volterra series analysis is introduced and shown to be an appropriate means to assess laser distortion. Finally, in section 7.5 the performance of SCM systems is assessed. Calculation of the intermodulation power spectral density is outlined and it is shown that laser distortion is conveniently quantified in terms of distortion coefficients that take into account important parameters such as laser operating point and channel frequency allocation. Laser intensity noise, receiver shot and thermal noise are then considered and optimum performance is identified. The subject of receiver design is also addressed and conditions of operation are identified where APD detectors may be advantageous. Examples are given that illustrate the application of the technique.

7.2   SCM optical systems

A diagram of a basic SCM system is shown in Figure 7.1. Digital or analogue information is upconverted to a narrow-band channel at high frequency to create subcarriers. Analogue (AM-VSB or FM) video signals used in CATV or satellite systems can be directly applied. The several electrical subcarriers are combined and

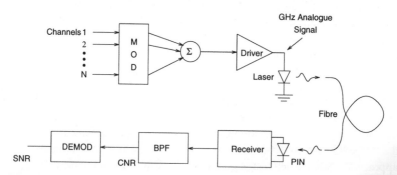

Figure 7.1 *Block diagram of a directly modulated subcarrier multiplexed system*

used to modulate the laser diode. The optical signal is then transmitted over the fibre and directly detected with a wide-band photodetector, prior to demodulation. This technique has inherent advantages. It uses conventional and established microwaves techniques for which components are commercially available taking advantage of the large bandwidth, exceeding 20 GHz, of lasers and detectors. Additionally, in multiple access applications, whilst the photodiode in each receiver detects all channels, only the desired narrow-band channel needs to be amplified and demodulated. This results in an increase of the receiver sensitivity as determined by the bandwidth of that particular channel. Also, the spectral characteristics of the source are unimportant so that multimode lasers can be used and temperature stabilisation is not critical. Finally, since individual channels are independent, SCM systems have great flexibility in allocating bandwidth and so can adapt rapidly to evolving changes.

However, two important sources of noise, relative intensity noise (RIN) and intermodulation distortion, originate at the laser diode. These will be discussed in the next section.

7.3 Intensity noise and nonlinear distortion

7.3.1 *Relative intensity noise*

Intensity noise embedded in a received signal directly degrades signal quality and thus system performance. At the most fundamental level, the origin of the laser intensity fluctuations lies in the quantum nature of the lasing process itself. Fluctuations arise from the spontaneous emission process and carrier-generation-recombination processes [19]. The emitted optical signal is assumed to exhibit noise, so that the time dependent photon density around the steady-state value P_0 is

$$P(t) = \langle P(t) \rangle + \delta P(t) \tag{7.1}$$

The intensity noise at a given frequency is characterised by the relative intensity noise (RIN) defined as

$$RIN = \frac{S_p(\omega)}{P_0^2} \tag{7.2}$$

where $S_P(\omega)$ is the spectral density of the random process $\delta P(t)$. This results in a noise spectral density, N_{RIN}, at the receiver given by

$$N_{RIN} = RIN \cdot I^2 \tag{7.3}$$

where I is the dc photocurrent.

In the semiclassical approach these fluctuations can be incorporated in the rate equations for the electron and photon densities by adding Langevin noise sources [19]. Using a small-signal analysis it is possible to obtain the relative intensity noise spectra. Figure 7.2 shows the RIN of a distributed feedback (DFB) buried heterostructure (BH) laser for several bias currents. The intensity noise is seen to peak at the resonance frequency ω_0 and to decrease with increase in laser power P_0. Below resonance and at low power levels RIN decreases according to $RIN \propto P_0^{-3}$; at higher bias levels RIN varies more slowly approaching the P_0^{-1} dependence.

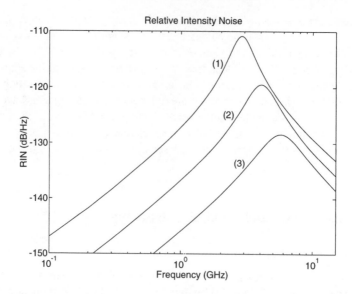

Figure 7.2 *Relative intensity noise spectra at several bias currents: (1) $I_0/I_{th} = 1.5$, (2) $I_0/I_{th} = 2$, (3) $I_0/I_{th} = 3$. Laser parameters taken from [20]*

7.3.2   RIN requirements

Consider analogue sinusoidal modulation with amplitude P_1 and average value P_0: $P(t) = P_0 + P_1 \sin(2\pi f t)$. The intensity noise, with power $\langle \delta P^2 \rangle$, is superimposed on the signal and a carrier-to-noise ratio for a modulation bandwidth B may be defined as

$$CNR = \frac{P_1^2/2}{\langle \delta P^2 \rangle} = \frac{m^2 P_0^2}{2\langle \delta P^2 \rangle} = \frac{m^2}{2 RIN \cdot B} \qquad (7.4)$$

with the optical modulation depth $m = P_1/P_0$. An improved CNR is obtained with increasing modulation depth m. The maximum value for m will be limited by nonlinear distortion (section 7.5). Assuming a high quality TV (AM) transmission with a signal-to-noise ratio of 56 dB (electrical dB after the photodetector, $SNR = 10^{56/10}$), $m = 0.08$ and $B = 5$ MHz this yields $RIN = -148$ dB/Hz. As seen above, RIN is bias-dependent and peaks at the resonance frequency. However, even at lower frequencies such a value for RIN is a stringent requirement for laser diodes. Moreover RIN can be increased significantly in the presence of reflections, from fibre discontinuities (fibre connectors/splices), back to the laser or by multiple reflections between fibre discontinuities [21]. These phenomena can increase RIN by 10–20 dB, and so in analogue intensity modulated systems care must be taken to minimise these effects.

7.3.3   Nonlinear distortion in semiconductor lasers

In the semiconductor laser, distortion can be divided into two types: static and intrinsic dynamic nonlinearity. Usually one type of nonlinearity is dominant depending on the frequency range of operation of the system.

For low modulation frequencies, below a few hundreds of megahertz, distortions are introduced mainly due to the imperfect linearity of the static light-power versus current (L-I) characteristic. This nonlinearity can be caused by leakage currents [19, 22] and is usually modelled using a power series expansion of the laser curve around the bias point [3]. The optical power of second-and third-order distortion products relative to the fundamental carrier are then proportional to (d^2L/dI^2) and (d^3L/dI^3), respectively. The leakage current effects can also be included in the rate equations by making the current through the active layer a nonlinear function of the total injected current. This requires accurate modelling of the electrical equivalent circuit of the laser chip [4, 23] but has the benefit that it leads to a complete model which includes both static and dynamic nonlinearity. With increasing modulation frequency the distortion increases rapidly caused by the nonlinear interaction between the photons and the injected carriers. This same nonlinear coupling is responsible for the relaxation oscillation resonance and so this distortion is denoted as intrinsic dynamic distortion or resonance distortion. This effect is well described by the rate equations.

Recently it has been shown that spatial hole burning in DFB lasers may play an important role in determining the nonlinear distortion at low frequencies [24]. Spatial hole burning is caused by the coupling between the DFB grating and the active region creating a non-uniform distribution of light intensity along the laser axis which is power-dependent [25]. This effect has been shown to be responsible not only for an additional nonlinearity of the laser curve but also for the gain and loss (threshold gain) above threshold becoming power dependent [26, 27]; that is modulation of the photon-lifetime, τ_p, occurs. The single-mode rate equations are then modified by the introduction of additional gain and loss suppression factors. This phenomenon, however, will not be considered in our analysis.

For directly modulated SCM systems operating in the high frequency region (> 1 GHz), such as FM systems, only the intrinsic dynamic nonlinearity of the laser needs to be considered since it is the dominant nonlinear mechanism. For directly modulated lasers at frequencies operated by AM systems (Table 7.1) leakage current and SHB should be considered by modification of the single-mode rate equations. Finally, if the L-I nonlinearity is eliminated and the dynamic resonance frequency is infinite, then the allowed modulation depth is limited by clipping. Below the threshold current the light output is zero and larger excursions in the modulation current produce distortion which limits the modulation depth per channel to about 5% for a carrier-to-noise ratio of 55 dB [28].

7.3.4   Linearity requirements

The laser dynamics are described by rate equations which are intrinsically nonlinear. In SCM systems where the laser is modulated by the sum of several subcarriers, mixing in the laser cavity occurs to form intermodulation products. Second-and third-order distortion products are generated by every combination of two and three input frequencies, respectively. The interference resulting from source nonlinearity then depends strongly on the number of channels and the distribution of channel frequencies. Let us consider transmission of three channels (Figure 7.3) with subcarrier frequencies f_1, f_2 and f_3. The third-order intermodulation distortion products

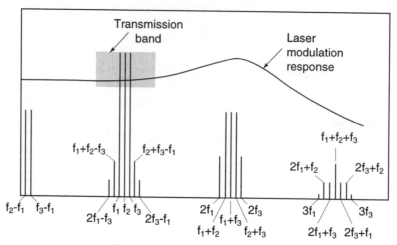

Figure 7.3 *Intermodulation products and harmonics generated by three-tone modulation of a laser diode*

(IMPs) at frequencies $f_i + f_j - f_k$ and $2f_i - f_j$ will certainly lie within the transmission band leading to interchannel interference. For a N channel system with uniform frequency spacing the number of IMPs $_rIM_{2\bar{1}}^N$ and $_rIM_{11\bar{1}}^N$ of type $2f_i - f_j$ and $f_i + f_j - f_k$, respectively, coincident with channel r are given by [29]

$$_rIM_{2\bar{1}}^N = \frac{1}{2}\left\{N - 2 - \frac{1}{2}\left[1 - (-1)^N(-1)^r\right]\right\} \tag{7.5}$$

$$_rIM_{11\bar{1}}^N = \frac{r}{2}(N - r + 1) + \frac{1}{4}\left[(N - 3)^2 - 5\right] - \frac{1}{8}\left[1 - (-1)^N\right](-1)^{N+r} \tag{7.6}$$

Figure 7.4 shows the total number of third-order IMPs as a function of channel number, each curve representing a different number of channels. For large N, $_rIM_{11\bar{1}}^N$ approaches the asymptotic value of $3N^2/8$ for the central carrier. For SCM systems occupying a bandwidth of more than one octave, the second-order nonlinear distortion will also have to be considered in which the most important terms are of type $f_i \pm f_j$.

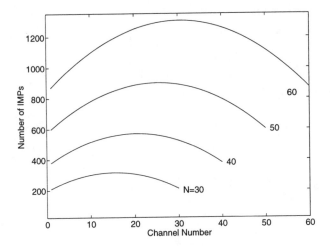

Figure 7.4 *Total number of third-order intermodulation products as a function of channel number with the number of channels, N as a parameter*

Table 7.1 compares the number of IMPs and the linearity requirements of typical demonstration SCM systems [30, 2, 31, 3, 32]. In the bidirectional digital system the advantage gained by reducing the bandwidth of each node outweighs the disadvantage of using high-frequency components required for time division multiplexing, even considering the expansion of bandwidth to 280 MHz in the upconversion process. In the FM system the baseband video signal is converted to an FM subcarrier with a bandwidth of typically 30–40 MHz. The increased bandwidth is exchanged, relative to the AM system, for a considerable signal-to-noise ratio (SNR) improvement so that the SNR at the demodulator output is much larger than the input CNR. A 16.5 dB CNR yields a high-quality signal with a 56 dB SNR [21].

Given the number of distortion products and the required carrier-to-noise ratio (CNR) the maximum magnitude of each type of product can be obtained. These numbers, which include a 10 dB margin, should be considered as an estimate because they do not include the frequency dependence of the distortion. For directly modulated AM systems the relatively small number of second-order products are generally more troublesome than the higher number of third-order products due to their higher amplitude. Also the high CNR required, 50 dB, compared to 17 dB of the bidirectional and FM systems, imposes greater restrictions on laser linearity.

7.4 Modelling laser resonance distortion

Our model for the semiconductor laser is based on single-mode rate equations with a nonlinear gain term [33], introduced phenomologically, which takes into account a

Table 7.1 *Linearity requirements of typical subcarrier multiplexed systems, after [4]*

	Bidirectional System (FSK)	FM Video	AM-VSB Video
# channels	8x180 Mb/s	60x30 MHz	42x4 MHz
Bandwidth (MHz)	280	30	4
Freq. Range (GHz)	2.5-5.0	2.4-4.8	0.05-0.40
Max. Product Count			
$f_i \pm f_j$	0	0	27
$f_i \pm f_j \pm f_k$	15	1276	576
CNR Required (dB)	17	17	50
Linearity Required (dB)			
$f_i \pm f_j$	–	–	-74
$f_i \pm f_j \pm f_k$	-39	-58	-88

number of nonlinear mechanisms; spectral hole burning [34] (symmetric nonlinear gain) and lateral carrier diffusion [35]. For simplicity of notation and computational purposes we use the normalised rate equations

$$\frac{dn}{dt} = j - n - (n - n_{0m})(1 - \varepsilon p)p \tag{7.7}$$

$$\frac{dp}{dt} = \gamma \left[\Gamma(n - n_{0m})(1 - \varepsilon p)p - p + \Gamma \beta n \right] \tag{7.8}$$

where n and p are the normalised electron and photon densities, respectively, Γ is the optical confinement factor, n_{0m} is the normalised electron density at which the net gain is zero, γ is the ratio of the spontaneous recombination lifetime τ_s to the photon lifetime τ_p, β is the fraction of the spontaneous emission coupled into the lasing mode and ε is the normalised gain compression factor; t is also normalised relative to τ_s.

Using a small-signal analysis of the rate equations, known as the perturbation technique, Lau and Yariv [14, 36] have calculated the two-tone third-order IMPs $2f_i - f_j$. Their results were later extended to InGaAsP lasers [15] and measurements have been given that show good agreement with theory [16]. Recently, further refinement of the formulae for the intermodulation distortion has been reported [37, 38, 39]. The results given are however restricted to two closely spaced microwave carriers. In the following section a more general analytical technique is discussed appropriate to the assessment of complex SCM optical systems.

7.4.1   *Volterra series analysis*

Signal distortion occurs in communication systems when the amplitude $|H(f)|$ and phase $\phi(f)$ of a linear network transfer function $H(f) = |H(f)|e^{i\phi(f)}$ are frequency dependent functions. This type of distortion is called "linear distortion". Besides this type of distortion other deviations can occur if the system possesses nonlinear elements. In this case, the system cannot be described by a single transfer function as in the linear case; instead the output, $y(t)$, is often expressed as a nonlinear function of the input $x(t)$, that is, $y(t) = T[x(t)]$. The resulting type of distortion is called "nonlinear distortion". In systems where the distortion is weak and frequency-dependent the nonlinearity is best represented by a Volterra functional series described as a "power series with memory" [40]. This analysis will be applied to assess laser distortion [41].

The semiconductor laser is then viewed as a nonlinear system where the output $p = p_0 + p(t)$, the normalised photon density, is some functional of its input $j = j_0 + j(t)$, the normalised current density,

$$p(t) = \sum_{n=1}^{\infty} p_n(t) \tag{7.9}$$

$$p_n(t) = \int_{-\infty}^{\infty} \cdots \int_{-\infty}^{\infty} h_n(\tau_1, ..., \tau_n) \prod_{r=1}^{n} j(t - \tau_r) d\tau_r \tag{7.10}$$

where p_0 and j_0 are the corresponding steady-state values and $h_n(t_1, ..., t_n)$ is the laser nonlinear impulse response of order n. The Fourier transform of $h_n(t_1, ..., t_n)$

$$H_n(f_1, ..., f_n) = \int_{-\infty}^{\infty} \cdots \int_{-\infty}^{\infty} h_n(\tau_1, ..., \tau_n) \prod_{r=1}^{n} e^{-i2\pi f_r \tau_r} d\tau_r \tag{7.11}$$

is the laser nonlinear transfer function of order n.

The method described in [40] to evaluate the transfer function is called the "probing" or "harmonic input" method because it assumes the input to be given by a sum of exponentials

$$j(t) = e^{i2\pi f_1 t} + e^{i2\pi f_2 t} + \cdots + e^{i2\pi f_n t} \tag{7.12}$$

where the $f_r, r = 1, 2, ..., n$ are linearly independent; that is, there are no rational numbers $m_1, m_2, \ldots, m_n$ (not all zero) such that $\sum_{r=1}^{n} m_r f_r = 0$. The laser is first excited by a single exponential the coefficient of which gives the first-order transfer function $H_1(f)$ associated with the photon density. Then a sum of two exponentials is applied yielding $H_2(f_1, f_2)$ in terms of $H_1(f_1)$. This procedure continues to the required order with one exponential being added at each step. Since third-order distortion products are of importance the transfer functions up to third-order have to be determined. General expressions for the transfer function $H_n(f_1, \ldots, f_n)$ of order n are obtained in [41].

In Figures 7.5 to 7.7 the analytical transfer functions are compared with the results of the numerical solution of the rate equations for a BH laser diode [20] when

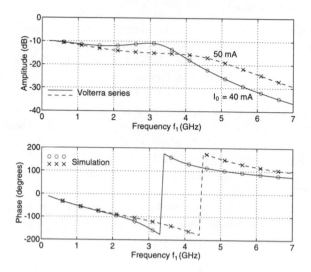

Figure 7.5 *First-order laser transfer function $H_1(f_1)$: comparison between analytical and simulation results when the laser is modulated by three carriers for two values of the laser bias current I_0 ($m(0) = 0.13$)*

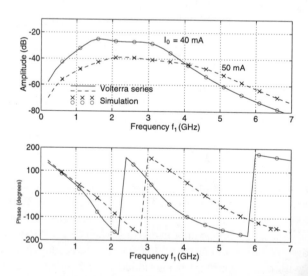

Figure 7.6 *Second-order laser transfer function $H_2(f_1, f_2)$: comparison between analytical and simulation results when the laser is modulated by three carriers for two values of the laser bias current I_0 ($f_2 = f_1 + 0.2$ GHz and $m(0) = 0.13$)*

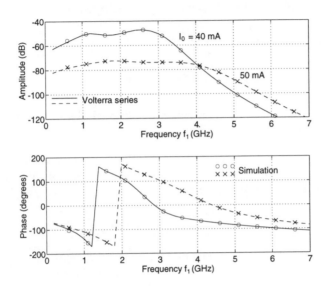

Figure 7.7 *Third-order laser transfer function* $H_3(f_1, f_2, -f_3)$: *comparison between analytical and simulation results when the laser is modulated by three carriers for two values of the laser bias current* I_0 ($f_2 = f_1 + 0.2\,\text{GHz}$, $f_3 = f_1 + 0.7\,\text{GHz}$ *and* $m(0) = 0.13$)

the laser is modulated with three sine waves. The input signal amplitude is defined in terms of the per-channel optical modulation depth m as

$$m(f) = \frac{|jH_1(f)|}{p_0} \tag{7.13}$$

The importance of this analysis is that it considers the general case of a sum of narrow-band signals.

7.4.2 Response to narrow-band signals

Let the input signal consist of a sum of N narrow-band channels centered at v_k

$$j(t) = \sum_{k=1}^{N} c_k(t)\cos(2\pi v_k t + \psi_k) - s_k(t)\sin(2\pi v_k t + \psi_k) \tag{7.14}$$

with a uniform distributed random phase ψ_k. The complex envelope of the nth-order output component at v

$$v = \sum_{r=1}^{n} v_{k_r} = \sum_{k=-N}^{N} m_k v_k \tag{7.15}$$

generated by intermodulation of the input signals centered at v_{k_r}, is given by [41]

$$
q_{nv}(t) = \frac{n!2^{-n+1}}{m_{-N}!\cdots m_N!} \int_{-\infty}^{\infty} \cdots \int_{-\infty}^{\infty} H_n(f_1 + v_{k_1}, \ldots, f_n + v_{k_n})
$$

$$
\times \prod_{r=1}^{n} Z_{k_r}(f_r)e^{i2\pi f_r t}df_r \tag{7.16}
$$

where $Z_k(f)$ is the Fourier transform of the complex envelope input signal

$$
z_k(t) = c_k(t) + is_k(t). \tag{7.17}
$$

We now assume $H_n(v_{k_1}, \ldots, v_{k_n})$ to be constant around any set of frequencies $v = (v_{k_1}, \ldots, v_{k_n})$. In this case the complex envelope of a particular intermodulation product falling at frequency v due to signals centered at $v_{k_1}, \ldots, v_{k_n}$ is [40]

$$
q_{nv}(t) = \frac{n!2^{-n+1}}{m_{-N}!\cdots m_N!}H_n(v) \prod_{r=1}^{n} z_{k_r}(t). \tag{7.18}
$$

If a particular frequency appears m_k times the product becomes

$$
\prod_{r=1}^{n} z_{k_r}(t) = \prod_{k=-N}^{N} z_k^{m_k}(t) \tag{7.19}
$$

and the waveform of the nth-order IMP falling at v is

$$
p_{nv}(t) = \frac{n!2^{-n+1}}{m_{-N}!\cdots m_N!}Re\left\{ H_n(v) \prod_{k=-N}^{N} e^{i2\pi vt}z_k^{m_k}(t)e^{im_k\psi_k} \right\} \tag{7.20}
$$

For the particular case of N input carriers at $v_k = f_k$

$$
j(t) = \frac{1}{2}\sum_{k=1}^{N}\left[j_k e^{i2\pi f_k t} + j_k^* e^{-i2\pi f_k t} \right] \tag{7.21}
$$

the complex envelope spectra of the carriers is

$$
Z_k(f) = j_k\delta(f)
$$

and the complex envelope amplitude, $A(f_{k_1}, \ldots, f_{k_n})$, of the nth-order IMP due to the carriers at $f_{k_1}, \ldots, f_{k_n}$ may be found from Equation 7.18

$$
A(f_{k_1}, \ldots, f_{k_n}) = \frac{n!j_{k_1}\cdots j_{k_n}}{2^{n-1}m_{-N}!\cdots m_N!}H_n(f_{k_1}, \ldots, f_{k_n}) \tag{7.22}
$$

$$
k_1, \ldots, k_n = \pm 1, \ldots, \pm N
$$

The amplitudes for three input carriers of typical second-and third-order IMPs are listed in Table 7.2. Figure 7.8 plots the analytical and simulation results for second-and third-order intermodulation products relative to the carrier. The results corresponding to a BH laser [20] are given for two values of the laser bias current. This analysis gives good results even for large modulation depths indicating that the third-order model of the laser is adequate. In [42] this same technique has been compared with a previously reported analysis, namely the perturbation technique.

Table 7.2 *Amplitude of third-order IMPs*

n	m_{-3}	m_{-2}	m_{-1}	m_1	m_2	m_3	Frequency	Amplitude
2	0	0	0	2	0	0	$2f_1$	$\frac{1}{2}j_1^2 H_2(f_1, f_1)$
2	0	0	0	1	1	0	$f_1 + f_2$	$j_1 j_2 H_2(f_1, f_2)$
3	0	1	0	2	0	0	$2f_1 - f_2$	$\frac{3}{4}j_1^2 j_2^* H_3(-f_1, f_2, f_2)$
3	0	0	1	0	2	0	$2f_2 - f_1$	$\frac{3}{4}j_1^* j_2^2 H_3(-f_1, f_2, f_2)$
3	1	0	0	1	1	0	$f_1 + f_2 - f_3$	$\frac{3}{2}j_1 j_2 j_3^* H_3(f_1, f_2, -f_3)$
3	0	1	0	1	0	1	$f_1 - f_2 + f_3$	$\frac{3}{2}j_1 j_2^* j_3 H_3(f_1, -f_2, f_3)$
3	0	0	1	0	1	1	$f_2 + f_3 - f_1$	$\frac{3}{2}j_1^* j_2 j_3 H_3(-f_1, f_2, f_3)$

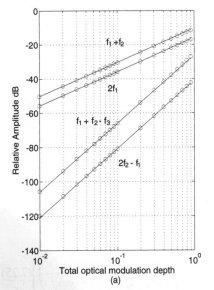

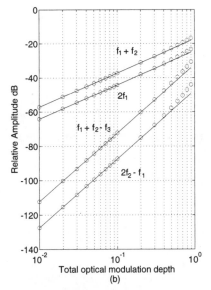

Figure 7.8 *Amplitude of second and third-order intermodulation products relative to the carrier at f_1 for a laser bias current of (a) $I_0 = 40\,mA$ and (b) $I_0 = 60\,mA$ ($f_1 = 2.1\,GHz$, $f_2 = 2.3\,GHz$ and $f_3 = 2.8\,GHz$)*

7.5   System performance assessment

In the previous section it was shown that the Volterra series provides a tractable analytical technique for the assessment of laser distortion. This technique will now be applied to assess the performance of SCM systems.

7.5.1   Intermodulation power spectral density

The power distribution of the multiplexed signal is affected by laser nonlinearity. In general the power of the intermodulation products spreads over a range of frequencies larger than the original channel bandwidth. The bandpass filter which selects the desired channel (Figure 7.1) then rejects some of this "intermodulation noise". To assess the impact of laser distortion on system performance it is therefore important to know how the power of the distortion products is distributed over frequency. The nth-order intermodulation power spectral density $G_{p_{n\nu}}(f)$ is calculated in [43] for the wideband-FM case assuming the transfer functions are constant over the channel bandwidth. The laser input current is considered to be a sum of N FM signals centered at ν_k

$$
\begin{aligned}
j(t) &= \sum_{k=1}^{N} A_k \cos[2\pi \nu_k t + \phi_k(t) + \psi_k] \\
&= \sum_{k=1}^{N} Re\left\{ z_k(t) e^{i(2\pi \nu_k t + \psi_k)} \right\}
\end{aligned}
\tag{7.23}
$$

the phase $\phi_k(t)$ being equal to the integral of the modulation signal $x_k(t)$

$$
\phi_k(t) = 2\pi \Delta f \int_0^t x_k(\xi)\, d\xi, \qquad \Delta f = \frac{\Delta f_{pp}}{2}
\tag{7.24}
$$

If p_{x_k} is the probability density function of the signal the final result is

$$
\begin{aligned}
G_{p_{n\nu}}(f) = {} & \frac{1}{4}\left[\frac{n!\,2^{-n+1}}{m_{-N}! \cdots m_n!} \right]^2 |H_n(\nu)|^2 \left[\prod_{k=-N}^{N} A_k^{2m_k} \right] \\
& \times \left\{ \frac{1}{m_{-N}\Delta f} p_{x_{-N}}\left(\frac{f-\nu}{m_{-N}\Delta f} \right) * \cdots * \frac{1}{m_N \Delta f} p_{x_N}\left(\frac{f-\nu}{m_N \Delta f} \right) \right. \\
& \left. + \frac{1}{m_{-N}\Delta f} p_{x_{-N}}\left(\frac{-f-\nu}{m_{-N}\Delta f} \right) * \cdots * \frac{1}{m_N \Delta f} p_{x_N}\left(\frac{-f-\nu}{m_N \Delta f} \right) \right\}
\end{aligned}
\tag{7.25}
$$

where Δf is half of the peak-to-peak frequency deviation and $*$ denotes convolution.

     Figure 7.9 illustrates the intermodulation power spectral density for the simple case when the modulating signal is uniformly distributed.

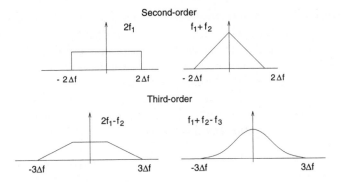

Figure 7.9 *Intermodulation power spectral density of second-and third-order IMPs; p_x uniformly distributed between -1 and 1*

7.5.2 Carrier-to-intermodulation ratio

In order to quantify the effect of intermodulation noise on the system performance we define the nth-order carrier-to-intermodulation ratio (CIR) for the rth channel located at frequency v, due to input channels at $f_{k_1}, ..., f_{k_n}$, as the ratio of the signal power S_0 to the nth-order intermodulation power at the output of the bandpass filter (BPF) (Figure 7.1), with transfer function $H_{BP}(f)$. CIR is then written as

$$CIR_{nr}(f_{k_1}, ..., f_{k_n}) = \frac{S_0}{\int_B |H_{BP}(f)|^2 G_{p_{nv}}(f)df} \qquad (7.26)$$

where the integral is evaluated over the filter bandwidth B. Since the channels are independent the total nth-order CIR for channel r is obtained by summation of all the IMPs power terms of order n falling at frequency v

$$CIR_{nr} = \frac{S_0}{\sum_k \int_B |H_{BP}(f)|^2 G_{p_{nv}}(f)df} \qquad (7.27)$$

where the summation over k includes all the distinct sets $\{k_1, \ldots, k_n\}$ such that

$$f_{k_1} + \cdots + f_{k_n} = m_{-N} f_{-N} + \cdots + m_N f_N = v.$$

Assuming that $H_{BP}(f)$ has a unit gain over its bandwidth, the CIR for the FM system considered previously [43] becomes

$$(CIR)_{nr} = \frac{\frac{1}{2}|H_1(v)|^2 A_r^2}{\frac{1}{2}B_{nm}^2 \left[\prod_{k=-N}^{N} A_k^{2m_k} \right] \alpha_n \sum_k \left| H_n(f_{k_1}, \cdots, f_{k_n}) \right|^2} \qquad (7.28)$$

where $B_{nm} = n! 2^{-n+1}/(m_{-N}! \cdots m_N!)$ and $\alpha_n(m_{-N}, \cdots, m_N)$ denotes the fraction of the intermodulation power relative to the signal power that is passed by the filter. The channels are uniformly spaced in frequency and so α_n does not depend on the specific frequencies which originate the distortion but only on the type of distortion product. Also since the channels are independent the total CIR for channel r is obtained by the summation of all the IMP power terms falling at frequency ν. Considering that the number of intermodulation products of type $f_i + f_j - f_k$ falling at a particular channel in a sequence of N equally spaced carriers, and for large N, increases as N^2 whereas $2f_i - f_j$ and $f_i - f_j$ increase as N, CIR can be conveniently expressed in terms of distortion coefficients D

$$(CIR)_r^{-1} = m^4 \left[D_{111} N^2 + D_{21} N \right] + m^2 \left[D_{11} N + D_2 \right] \tag{7.29}$$

The indices of the distortion coefficients D identify the order and type of distortion products and are

$$D_{111} = \left(\frac{3}{2} \right)^2 p_0^4 \frac{\alpha_{111}}{N^2} \sum_k \frac{\left| H_3(f_{k_1}, f_{k_2}, -f_{k_3}) \right|^2}{|H_1(\nu)|^6} \tag{7.30}$$

$$D_{21} = \left(\frac{3}{4} \right)^2 p_0^4 \frac{\alpha_{21}}{N} \sum_k \frac{\left| H_3(f_{k_1}, f_{k_2}, -f_{k_3}) \right|^2}{|H_1(\nu)|^6} \tag{7.31}$$

$$D_{11} = p_0^2 \frac{\alpha_{11}}{N} \sum_k \frac{\left| H_2(f_{k_1}, f_{k_2}) \right|^2}{|H_1(\nu)|^4} \tag{7.32}$$

$$D_2 = \left(\frac{1}{2} \right)^2 p_0^2 \alpha_2 \sum_k \frac{|H_2(f_k, f_k)|^2}{|H_1(\nu)|^4} \tag{7.33}$$

Expressions for α_n are given in [43].

Figure 7.10 shows the CIR for a 62-channel FM video system operating in the band of 2.7–5.2 GHz using a DFB-BH laser biased at 50 mA with a resonance at 9 GHz. Since the transmission bandwidth is limited to one octave the second-order IMPs and harmonics need not be considered. Also the third-order three-tone IMPs dominate for a large number of channels [1, 21]. At this bias point $D_{111} = 0.02$.

CIR is plotted as a function of the optical modulation depth per channel (m) for the first, middle and the last channels. In spite of the middle having the highest number of intermodulation products, 1365 compared to 900 for channel 62, the last channel is the most strongly affected by the intermodulation noise: $CIR_{62} < CIR_{31}$. Figure 7.11 shows the dependence of the distortion coefficient on the laser bias current I_0. The peak on this curve is due to the effect of the laser parasitics on the laser transfer function which causes a depression to occur at ≈ 5 GHz for $I_0 = 42$ mA. This means that if the channels have the same amplitude the modulation depth for the channels located at ≈ 5 GHz will be lower with a corresponding increase in the distortion coefficient.

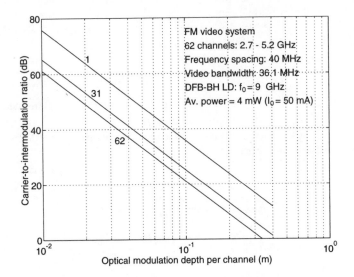

Figure 7.10 *Carrier-to-intermodulation ratio for channels 1, 31 and 62*

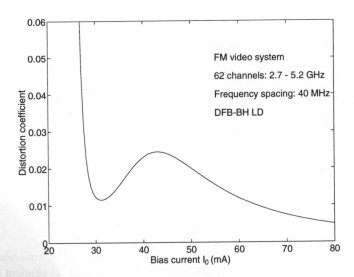

Figure 7.11 *Third-order distortion coefficient D_{111} for a DFB-BH laser*

7.5.3   System implications

Thus far, only contributions from laser nonlinearity have been considered. In the determination of the overall system CNR, optical intensity fluctuation noise, called relative intensity noise (RIN), receiver shot and thermal noise must also be taken into account. The CNR in the receiver may be expressed as:

$$CNR_r^{-1} = CNR_{TX_r}^{-1} + CNR_{RX_r}^{-1} \tag{7.34}$$

where CNR_{TX} is the ratio of the carrier power to the noise power generated by the laser diode and CNR_{RX} is the ratio of the carrier power to the noise power generated at the receiver. These ratios are relative to a specific channel r. For simplicity of notation we will drop the subscript r.

The ratios are given by

$$CNR_{TX}^{-1} = \left(\frac{m^2}{2BRIN} \right)^{-1} + CIR^{-1} \tag{7.35}$$

$$CNR_{RX} = \frac{\frac{1}{2}(mgI)^2}{(2eIg^2F + \langle I_r^2 \rangle)B} \tag{7.36}$$

with definitions as follows

    RIN: relative intensity noise
    g: APD gain
    $\langle I_r^2 \rangle$: receiver noise spectral noise density (pA/$\sqrt{\text{Hz}}$)
    I: primary dc photocurrent
    F=g^x: excess APD noise factor
    e: electronic charge
    B: signal bandwidth

The total CNR is then written as

$$CNR = \frac{\frac{1}{2}m^2I^2g^2}{\langle I_r^2 \rangle B + g^2I^2RINB + 2eBIg^2F + \frac{1}{2}g^2I^2 \{ m^6C_1 + m^4C_2 \}} \tag{7.37}$$

where

$$C_1 = D_{111}N^2 + D_{21}N \tag{7.38}$$
$$C_2 = D_{11}N + D_2 \tag{7.39}$$

CNR may be bivariately maximised in m and g for a given I, equivalent to balancing the total contributions of signal-independent and signal-dependent noise terms [44]. The maximum CNR for an optimum modulation depth, m_{opt}, for a given I is

$$CNR^{-1}(m_{opt}) = 3m_{opt}^4 \left[D_{111}N^2 + D_{21}N \right] + 2m_{opt}^2 [D_{11}N + D_2] \tag{7.40}$$

Conversely m_{opt} may be determined by specifying the required CNR

$$m_{opt}^2 = \frac{-C_2 + \sqrt{C_2^2 + 3C_1/CNR}}{3C_1}.$$

(7.41)

The optimum APD gain is given by the usual expression

$$g_{opt} = \left[\frac{\langle I_r^2 \rangle}{eIx}\right]^{1/(2+x)}$$

(7.42)

The maximum CNR for an optimum OMD and APD gain is then

$$CNR(m_{opt}, g_{opt}) = \frac{\frac{1}{2}m_{opt}^2 I^2 g_{opt}^2}{K_3 B \left\{\langle I_r^2 \rangle \left[1 + \frac{2}{x}\right] + g_{opt}^2 I^2 RIN\right\}}$$

(7.43)

where

$$K_3 = \frac{3m_{opt}^2 C_1 + C_2}{2m_{opt}^2 C_1 + C_2}$$

(7.44)

The necessary primary dc photocurrent to achieve $CNR(m_{opt}, g_{opt})$ is readily obtained by combination of Equations 7.42 and 7.43

$$I_{APD} = \left[\frac{ex}{\langle I_r^2 \rangle}\right]^{\frac{1}{1+x}} \left[\frac{CNR(m_{opt}, g_{opt}) K_3 B \langle I_r^2 \rangle (1 + 2/x)}{\frac{1}{2}m_{opt}^2 - RIN K_3 B}\right]^{\frac{1+x/2}{1+x}}$$

(7.45)

For a PIN receiver the gain and the excess noise factor are both unity and the required photocurrent to achieve $CNR(m_{opt})$ is then

$$I_{PIN} = \frac{e + \sqrt{e^2 + W\langle I_r^2 \rangle}}{W}$$

(7.46)

with

$$W = \frac{m_{opt}}{2CNR K_3 B} - RIN$$

(7.47)

where once again we have followed the notation of [44].

In practice we are interested in obtaining the receiver sensitivity for a desired CNR and a specific number of channels with the laser biased at a certain point. These last two parameters, number of channels and laser bias current, will determine the levels of distortion at the laser output and are included in the analysis through the distortion coefficients D. Once the system parameters are specified the distortion coefficients are determined and from these the optimum modulation depth for the desired CNR follows from Equation 7.41. Depending whether the system uses a PIN or an APD, the receiver sensitivity is readily obtained from Equations 7.45 or 7.46, respectively.

7.5.4 Receiver design considerations

It has been verified that the optimum modulation depth is determined by the inter-modulation distortion. Therefore, for a certain amount of distortion (given by the distortion coefficients), the photocurrent I must increase if higher values of CNR are required. Shot noise and RIN then start to dominate. The associated APD gain will reduce towards unity for a fixed circuit noise and APD receivers do not provide any advantage. A PIN diode is then the preferred choice. Even so APDs can improve performance for relatively low CNR particularly when circuit noise is dominant.

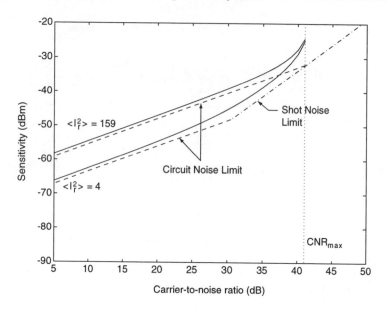

Figure 7.12 *Sensitivity as a function of carrier-to-noise ratio using a PIN detector for two circuit noise levels. Parameters are $R_0 = 0.8\,\mathrm{A/W}$, $N = 62$, $B = 36\,\mathrm{MHz}$, $RIN = -155\,\mathrm{dB/Hz}$ and $D_{111} = 0.02$*

Figures 7.12 and 7.13 show the performance obtained with a PIN detector and an APD, respectively. We have considered an APD with $x = 0.7$, responsivity of $R_0 = 0.8\,\mathrm{A/W}$ and two circuit noise levels corresponding to a tuned receiver with $\langle I_r^2 \rangle = 4\,\mathrm{pA^2/Hz}$ [45] and a 50 Ohm low-noise amplifier with a noise figure of 3 dB for which $\langle I_r^2 \rangle = 159\,\mathrm{pA^2/Hz}$. Other parameters are number of channels $N = 62$, $B = 36\,\mathrm{MHz}$, $RIN = -155\,\mathrm{dB/Hz}$ and third-order intermodulation $D_{111} = 0.02$ [31, 43]. The solid lines in Figure 7.12 represent the relation between CNR and the optical power for the optimum modulation depth, derived from equation 7.46 (values of the optimum modulation depth can be read from Figure 7.14 since m_{opt} is independent of the type of detector used). In this figure it is seen that when CNR ratios of less 25 dB are required the difference in sensitivity between the two circuits is approximately 8 dB, as both are operating in the circuit noise limit. Between 25

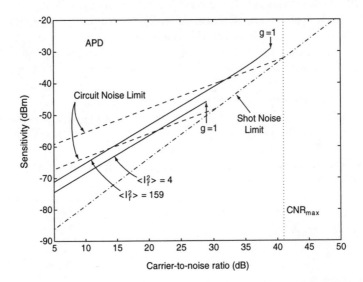

Figure 7.13 *Sensitivity as a function of carrier-to-noise ratio using an APD with the same assumptions as for Figure 7.12:* $R_0 = 0.8\,\text{A/W}$, $x = 0.7$, $N = 62$, $B = 36\,\text{MHz}$, $RIN = -155\,\text{dB/Hz}$ *and* $D_{111} = 0.02$

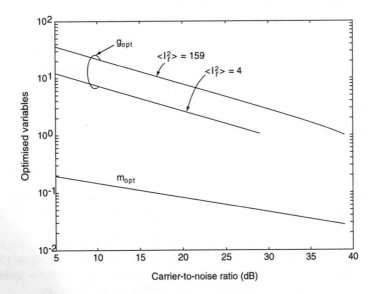

Figure 7.14 *Optimum avalanche gain and optimum modulation depth associated with the curves of Figure 7.13*

and 35 dB this difference is reduced and for CNR greater than 35 dB not only shot noise but also RIN and intermodulation distortion become important until a point is reached above which no transmission is possible.

Figure 7.13 shows the relation between CNR and the received optical power for an APD detector. The asymptotic values of the circuit noise limit for the two amplifier noise levels (using a PIN), as well as shot noise limit, are shown by dashed lines. The solid lines for the received optical power correspond to optimum avalanche gain and optimum modulation depth; these are shown in Figure 7.14. At low CNR the optimum avalanche gain is large, decreasing as the CNR increases and reaching a value of 1 (PIN) close to the point where the circuit and shot noise asymptotes intersect. The power margin relative to a PIN detector, for a CNR of approximately 16 dB, is 3.5 dB for the tuned receiver and 8 dB for the 50 Ohm amplifier. Thus, the APD can improve system performance by providing significant reduction in the required optical power for low CNR, such as for FM systems, but is ineffective for large CNR as required in AM systems.

7.6 Summary

Analytical techniques supporting the design and optimisation of practical SCM systems have been considered and the Volterra series method has been shown to be applicable to practical systems. Laser-related noise and distortion phenomena have been modelled and incorporated in the analysis and the linearity requirements of various systems have been assessed. Attainable performance for both AM and FM systems has been explored in terms of CNR and the resultant CNR and implications for receiver design for such systems have been considered. Overall we have found the Volterra series to represent a powerful and flexible tool applicable to system design and performance evaluation.

7.7 References

1 Darcie, T. E.: 'Subcarrier multiplexing for multiple–access lightwave networks', *J. Lightwave Technol.*, 1987, **LT-5**, pp. 1103–1110

2 Darcie, T. E., Iannone, P. P., Kasper, B. L., Talman, J. R., Burrus, C. A., and Baker, T. A.: 'Wide-band lightwave distribution system using subcarrier multiplexing', *J. Lightwave Technol.*, 1989, **7**, pp. 997–1005

3 Olshansky, R., Lanzisera, V. A., and Hill, P. M.: 'Subcarrier multiplexed lightwave systems for broadband distribution', *J. Lightwave Technol.*, 1989, **7**, pp. 1329–1342

4 Darcie, T. E.: 'Subcarrier multiplexing for lightwave networks and video distribution systems', *IEEE J. on Select. Areas in Commun.*, 1990, **8**, pp. 1240–1248

5 Su, C. B., Lanzisera, V., Olshansky, R., Powazinik, W., Meland, E., Schlafer, J., and Lauer, R. B.: '15 GHz direct modulation bandwidth of vapour-phase regrown 1.3 μm

InGaAsP buried-heterostructure lasers under CW operation at room temperature', *Electron. Lett.*, 1985, **21**, pp. 577–578

6 Olshansky, R., Powazinik, W., Hill, P., Lanzisera, V., and Lauer, R. B.: 'InGaAsP buried heterostructure laser with 22 GHz bandwidth and high modulation efficiency', *Electron. Lett.*, 1987, **23**, pp. 839–841

7 Lo, C. N.: 'A hybrid lightwave transmission system for subcarrier multiplexed video and digital B-ISDN services in the local loop', *J. Lightwave Technol.*, 1989, **7**, pp. 1839–1846

8 Olshansky, R., Lanzisera, V. A., and Hill, P.: 'Simultaneous transmission of 100 Mbit/s at baseband and 60 FM video channels for a wideband optical communication-network', *Electron. Lett.*, 1988, **24**, pp. 1234–1235

9 Gross, R., Olshansky, R., and Hill, P.: '20 channel coherent FSK system using subcarrier multiplexing', *IEEE Photonics Technol. Lett.*, 1989, **1**, pp. 224–226

10 Olshansky, R., Gross, R., and Schmidt, M.: 'Subcarrier multiplexed coherent lightwave system for video distribution', *IEEE J. on Select. Areas in Commun.*, 1990, **8**, pp. 1268–1275

11 Hill, P. M. and Olshansky, R.: '8 Gb/s subcarrier multiplexed coherent lightwave system', *IEEE Photonics Technol. Lett.*, 1991, **3**, pp. 764–766

12 Cheung, K. W., Liew, S. C., and Lo, C.: 'Experimental demonstration of multiwavelength optical network with microwave subcarriers', *Electron. Lett.*, 1989, **25**, pp. 381–383

13 Liew, S. C. and Cheung, K.: 'A broad-band optical network based on hierarchical multiplexing of wavelengths and RF subcarriers', *J. Lightwave Technol.*, 1989, **7**, pp. 1825–1838

14 Lau, K. Y. and Yariv, A.: 'Intermodulation distortion in a directly modulated semiconductor injection laser', *Appl. Phys. Lett.*, 1984, **45**, pp. 1034–1036

15 Darcie, T. E., Tucker, R. S., and Sullivan, G. J.: 'Intermodulation and harmonic distortion in InGaAsP lasers', *Electron. Lett.*, 1985, **21**, pp. 665–666

16 Iannone, P. and Darcie, T. E.: 'Multichannel intermodulation distortion in high-speed InGaAsP lasers', *Electron. Lett.*, 1987, **23**, pp. 1361–1362

17 O'Reilly, J. J. and Salgado, H. M.: 'Distortion analysis of semiconductor lasers: A caution', *Electron. Lett.*, 1991, **27**, pp. 946–947

18 Salgado, H. M. and O'Reilly, J. J.: 'Tractable models of laser distortion in subcarrier multiplexed optical systems': In *Digest of 3rd Bangor Symposium on Communications*, (Bangor, U.K.), 1991, pp. 23–25

19 Agrawal, G. P. and Dutta, N. K.: 'Long-wavelength semiconductor lasers', Electrical/Computer Science and Engineering Series, (Van Nostrand Reinhold, 1986)

20 Tucker, R. S. and Kaminow, I. P: 'High-frequency characteristics of directly modulated InGaAsP ridge waveguide and buried heterostructures', *J. Lightwave Technol.*, 1984, **LT-2**, pp. 385–393

21 Way, W. I.: 'Subcarrier multiplexed lightwave system design considerations for subscriber loop applications', *J. Lightwave Technol.*, 1989, **7**, pp. 1806–1818

22 Dutta, N. K., Wilt, P., and Nelson, R. J.: 'Analysis of leakage currents in 1.3 μm InGaAsP real-index-guided lasers', *J. Lightwave Technol.*, 1984, **LT-2**, pp. 201–208

23 Lin, M.-S., Wang, S.-Y., and Dutta, N. K.: 'Measurements and modeling of the harmonic distortion in InGaAsP distributed feedback lasers', *IEEE J. Quantum Electron.*, 1990, **26**, pp. 998–1004

24 Takemoto, A., Watanabe, H., Nakajima, Y., Sakakibara, Y., Kakimoto, S., Yamashita, J., Hatta, T., and Miyake, Y.: 'Distributed feedback laser diode and module for CATV systems', *IEEE J. on Select. Areas in Commun.*, 1990, **8**, pp. 1359–1364

25 Soda, H., Kotaki, Y., Ishikawa, H., and Imai, H.: 'Stability in single longitudinal mode operation in GaInAsP/InP phase-adjusted DFB lasers', *IEEE J. Quantum Electron.*, 1987, **QE-23**, pp. 804–814

26 Morthier, G., Libbrecht, F., David, K., Vankwikelberge, P., and Baets, R. G.: 'Theoretical investigation of the second-order harmonic distortion in the AM response of 1.55 μm F-P and DFB lasers', *IEEE J. Quantum Electron.*, 1991, **27**, pp. 1990–2002

27 Kawamura, H., Kamite, K., Yonetani, H., Ogita, S., Soda, H., and Ishikawa, H.: 'Effect of varying threshold gain on second-order intermodulation distortion in distributed feedback lasers', *Electron. Lett.*, 1990, **26**, pp. 1720–1721

28 Saleh, A. A. M.: 'Fundamental limit on number of channels in subcarrier multiplexed lightwave CATV system', *Electron. Lett.*, 1989, **25**, pp. 776–777

29 Westcott, R. J.: 'Investigation of multiple f.m./f.d.m. carriers through a satellite t.w.t operating near saturation', *Proc. IEEE*, 1967, **114**, pp. 726–740

30 Darcie, T. E., Iannone, P. P., Kasper, B. L., Talman, J. R., Burrus, C. A., and Baker, T. A.: 'Bidirectional multichannel 1.44 Gb/s lightwave system using subcarrier multiplexing', *Electron. Lett.*, 1988, **24**, pp. 649–650

31 Olshansky, R. and Lanzisera, V. A.: '60 channel FM video subcarrier multiplexed optical communication system', *Electron. Lett.*, 1987, **23**, pp. 1196–1198

32 Darcie, T. E. and Bodeep, G. E.: 'Lightwave subcarrier CATV transmission systems', *IEEE Trans. Microwave Theory Tech.*, 1990, **38**, pp. 524–533

33 Tucker, R. S.: 'High-speed modulation of semiconductor lasers', *J. Lightwave Technol.*, 1985, **LT-3**, pp. 1180–1192

34 Petermann, K.: 'Laser diode modulation and noise', Advances in Optoelectronics, (Kluwer Academic Publisher, 1988)

35 Tucker, R. S. and Pope, D. J.: 'Circuit modelling of diffusion on damping in a narrow-stripe semiconductor laser', *IEEE J. Quantum Electron.*, 1983, **QE-19**, pp. 1179–1183

36 Lau, K. Y. and Yariv, A.: 'Nonlinear distortions in the current modulation of non-self-pulsing and weakly self-pulsing GaAs/GaAlAs injection lasers', *Opt. Commun.*, 1980, **34**, pp. 424–428

37 Helms, Jochen: 'Intermodulation distortions of broad-band modulated lasers', *J. Lightwave Technol.*, 1992, **10**, pp. 1901–1906

38 Haldar, M. K., Kooi, P. S., Mendis, F. V. C., and Guan, Y. L.: 'Generalized perturbation analysis of distortion in semiconductor lasers', *J. Appl. Phys.*, 1992, **71**, pp. 1102–1108

39 Wang, J., Haldar, M. K., and Mendis, F.V. C.: 'Formula for two-carrier third-order intermodulation distortion in semiconductor laser diodes', *Electron. Lett.*, 1993, **29**, pp. 1341–1343

40 Bussgang, J. J., Ehrman, L., and Graham, J. W.: 'Analysis of nonlinear systems with multiple inputs', *Proc. IEEE*, 1974, **62**, pp. 1088–1119

41 Salgado, H. M. and O'Reilly, J. J.: 'Volterra series analysis of distortion in semiconductor laser diodes', *IEE Proc.-J*, 1991, **138**, pp. 379–382

42 Biswas, T. P. and McGee, W. F.: 'Volterra series analysis of semiconductor laser diode', *IEEE Photonics Technol. Lett.*, 1991, **3**, pp. 706–708

43 Salgado, H. M. and O'Reilly, J. J.: 'Performance assessment of broadcast FM optical subcarrier multiplexed systems', *IEE Proc.-J*, 1993, **140**, pp. 397–403

44 Walker, S. D., Li, M., Boucouvalas, A. C., Cunningham, D. G., and Coles, A. N.: 'Design techniques for subcarrier multiplexed broadcast optical networks', *IEEE J. on Select. Areas in Commun.*, 1990, **8**, pp. 1276–1284

45 Alameh, K. and Minasian, R. A.: 'Tuned optical receivers for microwave subcarrier multiplexed systems', *IEEE Trans. Microwave Theory Tech.*, 1990, **38**, pp. 546–551

Chapter 8

Optoelectronics for millimetre-wave radio over fibre systems

D. Wake

8.1 Introduction

Radio over fibre (RoF) transmission systems, characterised by having elements of free-space radio and optical fibre, are expected to find an increasing role in telecommunication networks over the next decade, due to their ability to provide operational benefits in a variety of applications. These can range from present day niche applications such as antenna remoting at satellite earth stations, to future mainstream applications such as a cordless or mobile periphery to an optical fibre network infrastructure. Future communication expectations are very likely to revolve around cordlessness or mobility while, in the same timeframe, major network operators are expected to expand their optical fibre infrastructure more and more into the access network. RoF systems provide good synergy between optics and radio, and are ideally placed, therefore, to allow an efficient means for these two seemingly disparate technologies to merge. This chapter sets out to cover some of the recent developments towards realising RoF systems operating at millimetre-wave frequencies, where the spectral resource that these future applications are likely to require can be obtained.

Some of the principal applications for RoF systems will be discussed in more detail in section 8.2, while in section 8.3 a selection of the possible system configurations are outlined. The development of a range of optoelectronic building blocks required in these system configurations is presented in section 8.4. System demonstrators, assembled using some of these components, are described in section 8.5 and conclusions arising from this work are made in section 8.6. First, however, a brief description of what RoF systems are, and why they provide operational benefits in telecommunications networks, is given below.

In essence, RoF systems utilise optical fibre transmission to deliver radio signals directly to a point of free-space radiation (antenna site). The layout of a simple system arrangement is shown in Figure 8.1.

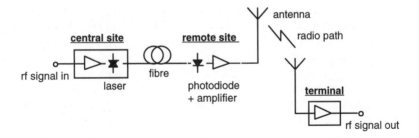

Figure 8.1 *Schematic drawing of simple one-way RoF system*

Depending on the application, the radio signals may be VHF, UHF, microwave or millimetre-wave. Optical fibre is an ideal transmission medium in this respect due to its low loss, high frequency and wide band capability. By delivering the radio signal directly, the optical fibre link avoids the necessity to generate the high frequency radio carrier at the antenna site, which is usually remote from easy access and also remote from a benign environment. For any transmission system, practical issues such as the size, weight, reliability, cost and power consumption of remote equipment are of critical importance. The principal advantage of RoF systems is the ability to concentrate most of the expensive, high frequency equipment at a centralised location, allowing the remaining equipment at the remote site to be simple, small size, light weight and low power consumption. This results in straightforward installation, low maintenance and a range of simplified options for electrical power provision. The centralisation of high frequency equipment also increases operational flexibility and the potential for re-use or sharing between a number of users. Just as importantly, the radiation frequency can be controlled centrally (away from the severe climatic variations suffered at remote sites) and therefore can be extremely stable.

8.2 Applications

Some of the applications for this technology that have been proposed include :
- satellite communications
  - antenna remoting and remote operation of satellite earth stations
- mobile radio communication networks
- dropwire replacement by radio
- broadband access by radio
- Multipoint Video Distribution Services (MVDS)
- Mobile Broadband System (MBS)
- vehicle communications and control
- radio LANs over optical networks

The first practical use of RoF technology comes from the area of satellite communications, for two main applications. The first application involves the remoting of antennas at satellite earth stations. The optical fibre links are short-range (0.1-1.0km) and operate at frequencies between 1GHz and 15GHz. They allow the centralisation of the high frequency equipment, which allows significant operational advantages to be gained such as quick and easy re-routing of traffic to different antennas via centralised switching. The second application is long-range (10-1000km) remoting of earth stations at VHF. Users can site their antenna many kilometres from the control area to improve satellite visibility or reduce interference with terrestrial systems, and cost savings are possible by removing control equipment from expensive metropolitan sites to allow more efficient use of high-cost premises.

The growth in cellular networks continues apace, with more than two million users connected in the UK at early 1994. As the number of users of these networks continues to rise, so the cell size has to be reduced to maintain a given number of users per cell. Microcell (around 1km diameter) and, in the longer term, picocell (around 100m diameter or less) architectures will be required to meet the demand. For such large scale deployment, cost will be a critical issue. RoF technology is likely to be the only means of driving the cost to a level low enough for such a massive infrastructure. It will enable the cheap, small, lightweight and low power consumption radio nodes that will be a prerequisite for such a network. RoF systems will also play an important role in the extension of coverage to areas with poor reception, such as underground stations, tunnels and also inside buildings.

In the nearer future there will be opportunities for the implementation of RoF systems in the local access network, where there is a need to provide telephony over radio [1]. Such cordless access networks must cost in against copper-fed radio systems. These systems are expected to appear within the next few years, operating either at CT2 (866MHz) or DECT (1890MHz) frequencies. The drive towards an information superhighway, when broadband image-based services start to gather momentum, will require radio in instances where cabling is impractical, where the customer requires cordlessness, or where rapid service deployment is required. Initially, frequency bands around 29GHz may be used for this purpose but, ultimately, frequencies around the 60GHz region will be required to provide adequate bandwidth.

Multipoint Video Distribution Services (MVDS) is a further opportunity to profit from the benefits of RoF technology. The European Community has recently allocated spectrum between 40.5GHz and 42.5GHz for this purpose. MVDS is a transmit-only service which can be used to serve areas the size of a small town with 30 or so channels of analogue FM TV. The coverage area is served by a transmitter located on a mast or tall building. Currently the transmitter consists of an array of directly modulated Gunn oscillators, each with its own horn antenna and heat pipe for frequency stabilisation. The rooftop hardware could be simplified considerably by using an optical fibre link to feed either a travelling wave tube or a solid-state amplifier at the transmit frequency. The weight and wind loading of the transmitter would be greatly reduced and the unit could be fed by a single optical fibre from a distance of several hundred metres.

The RACE (Research into Advanced Communications in Europe) Mobile Broadband System (MBS) is a futuristic system aimed at providing cordless connectivity between a mobile station and a fixed broadband network to satisfy demand for broadband services to pedestrians, trains and road vehicles. Frequencies between 62-66GHz have been allocated for this purpose and data rates up to 165Mb/s are envisaged. The system will use a high density of microcells, each up to a diameter of a few hundred metres. Optical fibre links will be used to interconnect these cells with the fixed network, so that transceiver equipment can be simple and low cost. Again, this is essential for systems with high density coverage.

Millimetre-wave radio LANs could be implemented in a very similar manner. Portable computers are becoming more powerful and more widely used, and wireless connectivity will be widespread in a few years time. A broadband wireless connection between workstations and a central file server could be realised in a simple way using optical fibre installed in each office, attached to a small, low-cost antenna unit.

Vehicle communication and control is another potential application area for RoF technology. A Europe-wide system is envisaged, and frequencies between 63-64GHz and 76-77GHz have been set aside. Such a network will require the distribution of millimetre-wave signals with low data rate information along major European roads. RoF systems could accomplish this with relatively low cost which, again, will be essential for such a large scale and extensive network.

8.3 System configurations

The first few systems referred to above operate at VHF, UHF or microwave frequencies and use a technique known as subcarrier multiplex (SCM) for the transmission of radio signals over the optical links. SCM is a technique in which a conventional frequency division multiplexed radio signal is applied directly to an intensity-modulated laser. The radio signal can consist of a free mixture of analogue or digital data channels. The main issues for this type of system are related to the performance of the laser, in terms of frequency response, noise and linearity.

Components and subsystems operating at frequencies up to around 18GHz can be obtained commercially. However, as discussed in section 8.2, RoF systems operating in the millimetre-wave bands are of interest for applications such as MVDS and MBS. The advantages of working in these bands stem from factors such as the availability of large chunks of spectrum, good link budgets due to high antenna gain, small physical antenna size and good frequency re-use resulting from high propagation losses beyond line-of-sight paths and also from atmospheric attenuation. For this type of system to be deployed on any significant scale, a low-cost method of generating and detecting optical millimetre-wave signals must be found. Presently, direct modulation of laser diodes is not feasible at millimetre-wave frequencies. Although optical modulators are now commercially available working at frequencies up to 50GHz, they are expensive and require high drive voltages. The generation of optical millimetre-wave signals using cheaper, low frequency optoelectronic components is currently the subject of active research. Methods such as resonant enhancement of laser response [2], harmonic generation [3] and coherent optical mixing [4] have been explored. Resonant enhancement of modulation response has been demonstrated using external cavity lasers, where enhancement takes place at frequencies corresponding to multiples of the round-trip time of the cavity. Harmonic generation techniques allow low-cost, relatively low-frequency optoelectronic components to be used as part of millimetre-wave systems. In all such components, use is made of the inherent nonlinearity of the optical response to the electrical input signal. Both lasers and modulators have been used for this purpose and useful millimetre-wave generation has been achieved using high order harmonics. Coherent optical mixing can also be used to generate millimetre-wave signals. If two coherent optical carriers are incident on the same photodiode, then there will be a component of the signal at the output of the photodiode at the difference frequency between the two carriers. For example, a wavelength difference of 0.5nm at a centre wavelength of 1550nm will produce a beat signal at a frequency of a little over 60GHz. In general, this technique suffers from signal purity and stability problems. Harmonic generation, conversely, produces signals with purity derived from the drive oscillator, which can be synthesised to produce output signals with sub-Hz linewidth. Some of the optoelectronic components developed for these various generation techniques will be discussed in the next section.

Ultra high speed photodiodes are now commercially available, but are expensive and have low efficiency. Millimetre-wave photodetection with high efficiency and potentially low-cost manufacture is another subject of active research. Examples include high efficiency photodiodes, photodiodes monolithically integrated with optical preamplifiers and heterojunction bipolar phototransistors (photoHBTs). Again, these components will be discussed in more detail in the next section.

A block diagram of a simple millimetre-wave transceiver configuration is shown in Figure 8.2. The incoming optical signal is detected and one of the sidebands is selected for amplification and transmission to the mobile station (not shown). The filter also serves to provide the millimetre-wave carrier for the return path. The received signal from the mobile is amplified before downconversion to IF for the return path of the optical link. This configuration simplifies the return path optoelectronics, since a relatively low frequency, directly modulated laser can be used.

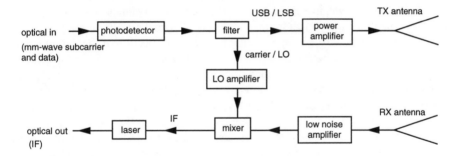

Figure 8.2 *Simple millimetre-wave transceiver configuration*

Alternative transceiver configurations are conceivable that make greater use of optoelectronic components such as optically injection-locked oscillators and optoelectronic mixers. Optically injection-locked oscillators can be important components in RoF systems because they can generate high output powers with high purity and stability from a relatively cheap oscillator. In other words, the oscillator can be designed for high power and low cost, without being constrained by purity and stability requirements. In configurations that make use of this type of component, a locking signal is transmitted over an optical link to transfer the purity of a reference source located centrally in a benign environment to the remote oscillator. Systems can be designed using this component in which the data is transmitted in baseband form using cheap optical sources. The locking signal can be transmitted separately, possibly using another fibre, in a configuration shown in Figure 8.3.

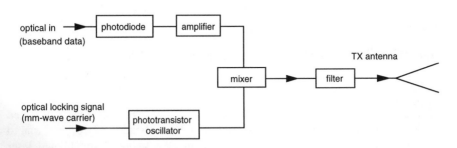

Figure 8.3 *System configuration using optically injection-locked oscillator (send path only)*

Optoelectronic mixers are also important building blocks in some RoF system configurations [5], where they could be used, for example, to upconvert baseband data on an optical carrier using an electrical LO derived in a similar way to the configuration shown in Figure 8.3.   This has the advantage of not having to convert the baseband optical input to an electrical signal prior to mixing.

Another area of active research is the further simplification of the radio node by the reduction of dc power consumption. Although RoF systems have the effect of allowing a relatively simple radio node, the development of very low power consumption optoelectronic components will provide further operational benefits. Provision of dc power at remote sites is a major issue in the deployment of any system. Power consumption at the remote radio site can therefore be an issue of critical importance. Provision of remote mains power takes considerable effort and expense, and can be a major factor in planning network infrastructure. Using a copper power cable from a network node is also problematic. Firstly it dilutes the advantages of an all-fibre network, possibly leading to problems with power surges and transients and, secondly, efficient power transfer requires high voltage levels, which brings problems from a safety standpoint. If the power consumption at the remote radio site can be reduced to a level where solar power with battery backup is a viable option, then many of the powering issues can be ignored. Efficient photodiodes, low current lasers and high efficiency amplifiers have been combined at BT Laboratories to create an experimental two-way radio node operating at 2.5GHz, with a total dc power consumption of only 40mW. The transmit power was around 1mW, which could give a range of up to a few hundred metres, depending on the application. This node demonstrated the feasibility of using solar power, with battery back-up, to remove the requirement for external power provision.

The next section looks at some of the options for millimetre-wave optoelectronic components, for use in the RoF system configurations discussed above. In general, these components fall into two categories. The first are novel components that are being designed and investigated in a R&D environment, which are therefore not available commercially at present. The second are components that are available commercially, but are being used in novel ways.

8.4 Millimetre-wave optoelectronic components

8.4.1 FM laser

One method of generating millimetre-wave signals makes use of a laser designed such that its optical frequency can be modulated by application of a drive signal to one of its terminals [3]. The optical spectrum of a frequency-modulated laser contains lines spaced by the drive frequency, and millimetre-wave signals are generated by the beating between widely spaced sidebands on a photodetector. A pure FM signal has constant intensity, and would not induce any photocurrent at harmonics of the drive frequency. However, if the light is propagated over dispersive optical fibre, then the relative phasing of the optical sidebands is altered, and the light acquires intensity fluctuations at harmonics of the laser drive frequency.

The process of modifying an optical FM spectrum through dispersion in single mode fibre can be analysed theoretically, and an expression derived for the intensity modulation depth of each harmonic at the output of the photodetector. The modulation depth M_p, of the pth harmonic is defined as the ratio of the amplitude of the alternating photocurrent at the pth harmonic to the dc photocurrent. The resulting expression for the modulation depth is [3]

$$M_p = |\, 2\, J_p \,(2\, \beta \sin(p\phi))\,| \tag{8.1}$$

where J_p is the pth Bessel function of the first kind, β is the FM index of the laser, and ϕ is an angle characterising the fibre dispersion given by

$$\phi = -\frac{\omega^2}{4\pi}\frac{D\lambda^2}{c}\,z \tag{8.2}$$

where λ is the free space wavelength of the laser, c is the speed of light, z is the fibre length, D is the fibre group velocity dispersion parameter, and ω is the angular frequency of the drive signal applied to the laser. The greatest modulation depth that can be obtained for the pth harmonic is therefore equal to twice the greatest value of the pth Bessel function ($p \geq 1$), and occurs when

$$2\,\beta\sin(p\phi) = j'_{p,1} \tag{8.3}$$

where $j'_{p,1}$ is the first zero of the derivative of the pth Bessel function. Thus to obtain the greatest modulation depth of the pth harmonic at the receiver it is necessary to adjust the FM index β to an optimum value, which is itself minimised if the fibre length is chosen such that $p\phi=\pi/2$. The theoretical dependence of the modulation depth of the 5th, 10th and 15th harmonics on the FM index is shown in Figure 8.4, where a value of $\phi=0.085$ has been assumed. It is possible to achieve up to 60% modulation depth for the 10th harmonic, and the corresponding values fall off quite slowly for higher harmonics. Therefore, the technique of FM-IM conversion offers the possibility of providing an efficient upconversion process.

This theory was evaluated experimentally using a three section, BH DFB laser with an FM efficiency of approximately 1GHz/mA [6] as the source. A 4GHz drive signal was applied to the centre section, and the output of the laser was launched into 12.5km of conventional step index fibre. A high-speed photodetector was employed as the optical receiver. A spectrum analyser was used to measure the magnitudes of the photocurrent at harmonics of the drive frequency applied to the laser, and the photodetector was calibrated so that the modulation depth of each harmonic could be calculated.

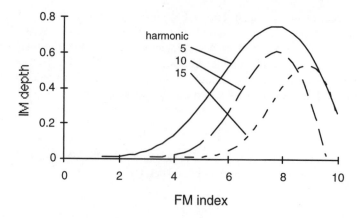

Figure 8.4   *Intensity modulation depth against FM index*

Values of modulation depth were measured by this technique as a function of the drive current applied to the laser. While the qualitative agreement with the theoretical results was good, the actual modulation depth of the 10th harmonic was around 13%, compared with the theoretical optimum of 60%. This discrepancy has been shown to be due to the large intensity modulation present at the output of the laser itself, which significantly alters the optical spectrum from that of a pure FM signal.

8.4.2 Mach-Zehnder modulator

The FM-IM conversion technique described above has been shown to be an effective method of producing millimetre-wave signals. However, it relies on the availability of a laser capable of wide optical frequency deviation at microwave rates; the peak-to-peak frequency deviation must be at least as great as the millimetre-wave frequency it is required to generate. This has been achieved through the use of a three-contact laser, although even with this device only 13% modulation depth was obtained. It would, therefore, be desirable to find an alternative source which could improve the modulation depth, and hence the overall efficiency of the conversion process.

Possible alternatives are provided by various types of external optical modulator. For example, the theory derived above would also be suitable for the case of an external phase modulator with cw light input. The advantage of this method is that the intensity modulation inherent with the directly modulated laser is not present, which means that the theoretical values of intensity modulation depth are realisable in practice. This has been verified experimentally, although is not reproduced here. The problem that still exists with both of these methods is the requirement for dispersion. Although the techniques are tolerant to large variations in fibre length, there will be circumstances where the fibre length is such that the intensity modulation fluctuations will fade.

One possibility for generating high order harmonics without recourse to dispersive fibre is to make use of the non-linear transfer characteristic of a Mach-Zehnder amplitude modulator. If a Mach-Zehnder modulator is driven with a sinusoidal input voltage $V\sin(t)$, then the resulting normalised intensity directly at the output of the device is given by

$$i(t) = \frac{1}{2}\,(1 + \cos\,(\Delta\beta\sin\,(\omega t) + \Delta\xi\,)) \tag{8.4}$$

where $\Delta\beta = \pi V/V_{SW}$ and V_{SW} is the voltage change required to switch from completely transmitting to completely blocking the light. The phase angle $\Delta\xi$ describes the operating point of the device which is controlled by a dc bias voltage. The cosinusoidal transfer characteristic gives rise to harmonics of the fundamental modulating frequency in the output intensity, $i(t)$. The modulation depth for each harmonic is slightly more awkward to define than for a phase- or frequency-modulated signal as the mean optical power transmitted through the modulator (and hence the dc photocurrent) depends on the bias point and drive voltage ($\Delta\xi$ and $\Delta\beta$ respectively). In the following the modulation depth is defined relative to the mean photocurrent when the modulator is biased at its half power point, such that $\Delta\xi=\pi/2$. The modulation depth is then given by

$$M_p\,(\Delta\beta) = \begin{cases} 2j\sin(\Delta\xi)\,J_p(\Delta\beta\,), & p = 1,3,5... \\ 2\cos(\Delta\xi)\,J_p(\Delta\beta\,), & p = 2,4,6... \end{cases} \tag{8.5}$$

The modulation depths that are achievable using the Mach-Zehnder modulator are therefore comparable to those achievable in dispersive FM or PM to IM conversion. To generate the odd harmonics with greatest efficiency requires the modulator to be biased such that $\Delta\xi=\pi/2$ (about which point it has an odd transfer characteristic), whereas to generate the even harmonics efficiently requires $\Delta\xi=0$ (about which it has an even characteristic).

In order to confirm the above theory, a lithium niobate Mach-Zehnder amplitude modulator with a switching voltage of 5V was driven at 1GHz, and harmonics of the intensity variations were observed at frequencies up to 5GHz. The theoretical modulation depth as a function of $\Delta\beta$ and the measured values as a function of drive voltage are shown in Figure 8.5. The close agreement between the theoretical and measured vaules confirms the theory and demonstrates that it is possible to achieve large modulation depth using this technique.

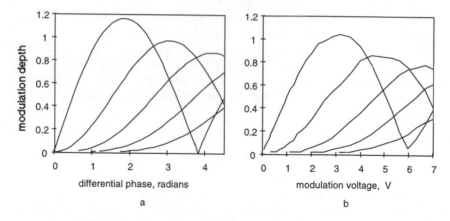

Figure 8.5 *Modulation depth achievable with Mach Zehnder modulator for first five harmonics: (a) theory ;   (b) experiment*

In order to generate millimetre-wave harmonics it is necessary to increase both the fundamental modulating frequency as well as the drive voltage over and above those in this experiment. Modulators are available commercially with a specification that would enable generation of 60GHz with +33dBm (2W) of RF power applied at 15GHz. Further reductions in the drive power should be possible through the use of a specialised electrode structrure designed for narrow-band operation at the modulating frequency.

Millimetre-wave signals at 36GHz have been generated experimentally using a Mach-Zehnder modulator driven at 18GHz (2nd harmonic) in a suppressed carrier scheme [7].   The resulting optical spectrum consists of two optical sudebands, which can be separated using optical filtering. Data can be introduced to just one of these sidebands, which is then recombined with the unmodulated sideband prior to transmission.   This arrangement provides an elegant method of adding data to the system, without incurring serious dispersion penalties when operating away from the dispersion minimum of the fibre [8]. This topic will be discussed in more detail elsewhere in this book.

8.4.3 Electroabsorption modulator

The electroabsorption (EA) modulator is a promising alternative to the Mach-Zehnder modulator for harmonic generation, with the potential for relaxed drive power requirements. The EA modulator is a semiconductor optical waveguide device in which the degree of optical absorption can be controlled by an applied voltage.   It is therefore an intensity modulator, like the MZ, but works on the principle of electric field induced absorption changes at photon energies close to the band edge of a semiconductor (the Franz-Keldysh effect in bulk material, or the quantum confined Stark effect in quantum well material). Devices have been fabricated in a variety of forms - discrete, integrated (with DFB laser), bulk absorption layer and multiquantum well (MQW) absorption layer.

The transfer characteristic of this type of device is very nonlinear, and has been used to good effect to generate picosecond optical pulses from a sinusoidal drive voltage. Electroabsorption modulators have also been assessed for harmonic generation and compared to the Mach-Zehnder in terms of output power and required drive voltage [9].

This MQW device was fabricated with an electroabsorption layer that consisted of 17 periods of 95Å InGaAsP wells (λ=1.55µm) and 55Å InGaAsP barriers (λ=1.1µm). A schematic drawing is shown in Figure 8.6.

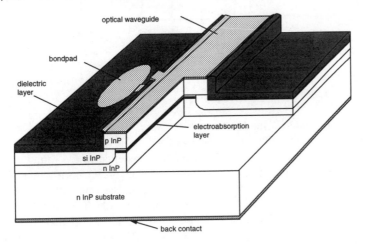

Figure 8.6 *Schematic drawing of electroabsorption modulator structure*

The chip consists of a reverse-biased PIN junction embedded in an iron-doped (semi-insulating) current blocking structure, which ensures that the modulating voltage is applied efficiently across the depleted electroabsorption layer. The optimum choice of chip length is determined by conflicting requirements for low capacitance (short length) and high modulation depth (long length). Chips 300 - 350µm long were chosen as a suitable compromise, and packaged in a fixture with an SMA connector for modulating voltage input. A small-signal bandwidth of over 10GHz is achieved by this means. The absorption characteristic for a typical device is shown in Figure 8.7, where an insertion loss of 10.8dB at 0V is apparent.

Harmonic generation using the electroabsorption modulator was modelled by fitting a spline curve to the transfer function of Figure 8.7, and calculating the Fourier series amplitudes of the transmitted light intensity that results from applying a sinusoidal drive voltage. Results were normalised for an optical input power of 1mW, and ideal photodiode characteristics were assumed for conversion to rf output power. The actual modulator insertion loss was used in the calculations.

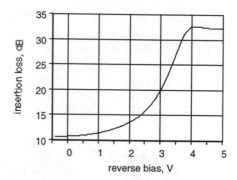

Figure 8.7   *Transfer function of electroabsorption modulator at a wavelength of 1560nm*

The Mach-Zehnder modulator which was used for comparison was a commercial LiNbO$_3$ device with a dc V_π of 8V, an insertion loss of 9.5dB, and a 3dB electrical bandwidth of 8GHz. This was modelled in the same way as the electroabsorption modulator. The effective value of V_π is different from the figure specified as a result of the internal impedance transformation from the system impedance (50Ω) to the modulator transmission line impedance (22Ω). For this particular modulator, the effective value of V_π was calculated to be 15V.

The electrical harmonic power for each modulator was measured using a spectrum analyser to monitor the output of a high speed photodiode. The drive frequency was set to 2.5GHz so that the results were not affected by the limited electrical frequency response of the modulators. Optical input power was set to 1mW, to coincide with the power used for the modelled results.

Figure 8.8 shows modelled and experimental results for each type of modulator, for different levels of rf drive power measured in a 50Ω load (P_{dr}). The plots show levels of rf output power for the first five harmonics of the drive frequency. For each harmonic, the dc bias voltage was set to optimise the rf output power. Figures 8.8(a), (c) and (e) show modelled results for drive powers of 16dBm, 22dBm, and 28dBm respectively. Figures 8.8(b) and (d) show experimental results for drive powers of 16dBm and 22dBm respectively.

Experimental results at 28dBm drive power were not obtained, but the close agreement between modelled and experimental results at the lower drive powers gives some confidence for the accuracy for Figure 8.8(e). Electroabsorption modulators can be seen to be particularly effective at moderate drive power levels and for higher order harmonics, where a significant advantage can be obtained over the Mach-Zehnder device. This advantage can be even greater in systems that use optical power amplification at the source. For any given harmonic, high modulation depth can be obtained with the EA modulator for relatively low levels of drive power. High modulation depth enables optical power amplifiers to realise high gain before saturation effects limit the output power. In our experiments, an optical amplifier with around 30dB gain and 15dBm saturated output power gave a third harmonic power level of 9dBm when corrected for photodiode response. This was achieved for a modulator drive power of only 22dBm.

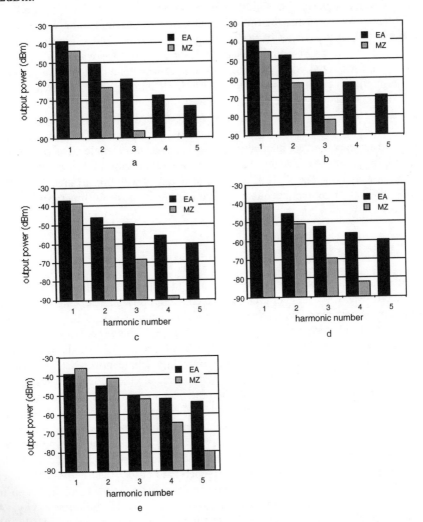

Figure 8.8   *Comparison between EA and MZ modulators for harmonic generation: (a) modelling; (b) experiment; $P_{dr} = 16dBm$;*
(c) modelling; (d) experiment; $P_{dr} = 22dBm$;
(e) modelling; $P_{dr} = 28dBm$

Generation of signals at 30GHz has been achieved using this EA modulator. The device was driven at 10GHz, with an electrical drive power of 22dBm and an optical input power of 1.5dBm. A photodiode with a responsivity of 0.2 A/W at 30GHz was used to produce a 3rd harmonic output signal power of -68dBm, which is equivalent to a power of -52dBm when corrected for photodiode response. As expected, the purity of this signal was determined by that of the drive oscillator (sub-Hz linewidth). Higher power levels from the photodiode can be obtained by increasing the optical input power to the modulator, or by optical amplification at the output of the modulator. Using the optical power amplifier described above, an rf output power at 30GHz of 14dBm was obtained (after correction for photodiode response). Millimetre-wave photodetectors with 100% quantum efficiency and no saturation at very high optical power are obviously not possible, making this corrected output power unobtainable in practice. However, quantum efficiency greater than 50% has been achieved in millimetre-wave photodiodes (see section 8.4.5) and optical loss in the form of network splits is likely to be encountered in practical systems. As an example, if a photodiode with 50% quantum efficiency was used in a system with an 8-way split, an rf output power at 30GHz of -10dBm could be obtained with this arrangement. This power level is more than adequate for many applications of millimetre-wave RoF systems.

8.4.4 Dual-mode laser

Optical mixing is an alternative approach to using harmonic techniques for millimetre-wave signal generation. If two coherent optical carriers are incident on a high-frequency photodiode, then there will be a component of the signal at the output of the photodiode at the difference frequency between the two carriers. Optical mixing, or heterodyning, is capable of producing signals with 100% intensity modulation depth, but suffers from high phase noise if two standard single-mode semiconductor lasers are used to generate the optical carriers. If the two optical modes are uncorrelated, then the beat signal linewidth is of the order of the combined optical-mode linewidth. For this type of laser, it is typically between 10 and 100MHz. The beat signal phase noise can be improved considerably by several methods, such as reducing laser linewidth using external cavities and electrical or optical feedback techniques to control and lock the frequency or phase. A more elegant approach is to ensure that the optical modes become correlated so that the beat noise is cancelled. The logical first step to achieving this correlation is to have the two optical modes in the same cavity. A dual-mode laser has been developed to test the viability of this approach [10].

Conventional distributed feedback (DFB) lasers are designed to operate with single longitudinal mode output. This is usually accomplished by a phase shift in the Bragg grating, or by the use of asymmetric facet coatings. Without this phase shift, an anti-reflection coated DFB laser would operate with two modes. The dual-mode laser therefore is designed without this phase shift, which ensures that no oscillation occurs at the Bragg frequency. Degeneracy gives rise to two modes, one either side of the Bragg frequency, separated by a stop-band. The design of the device structure is then modified to give the required mode separation. The major parameter governing the mode separation in a DFB laser is the grating strength coefficient, κ. Design data have been obtained, based on the expression for the FWHM bandwidth of a grating filter, which is given by [11]

$$\Delta f \approx \frac{\sqrt{2}\; c\; \kappa}{\pi\; n_e\; \tanh(\kappa L)} \tag{8.6}$$

where Δf is the mode frequency separation, L is the device length, and n_e is the effective refractive index without the grating. A κ value of 9cm^{-1} was chosen to give the required mode separation of around 60GHz for a 2mm long device (standard DFB structures are designed with typical κ values of around 30cm^{-1}). Buried heterostructure lasers were fabricated to this specification with a multi-section top electrode configuration.

The optical output from the lasers was observed on an optical spectrum analyser; a sample optical spectrum is shown in Figure 8.9(a). The mode separation for this device is 0.48nm. The electrical beat spectrum from each device was displayed on an rf spectrum analyser via a preselected external mixer and a high frequency photodiode. As expected, beat signals were observed corresponding to the optical mode separations measured using the optical spectrum analyser. An example is shown in Figure 8.9(b), which corresponds to the optical spectrum shown in Figure 8.9(a), where the beat frequency is around 57GHz. The linewidth of this beat signal is around 150MHz showing that the optical modes are not correlated in phase. Direct electrical injection at a subharmonic of the beat frequency was attempted and Figure 8.9(c) shows the result. Electrical modulation was applied to one of the laser contacts at a frequency of 6.3GHz (9th subharmonic) and a power level of 15dBm. Phase-locking has produced a signal with high purity (linewidth less than 10Hz and residual phase noise of -77dBc/Hz at 10kHz offset). The modulation depth of this signal is around 10-20%, as a result of the strong component at the fundamental frequency. This is an unavoidable consequence of subharmonic locking. However, the advantages of subharmonic locking (low frequency oscillator can be used and greatly relaxed high frequency device packaging) more than compensate for a reduced modulation depth.

The unlocked beat signal frequency can be changed over a range of around 1GHz with single contact current tuning, and the locking bandwidth of the locked signal is around 500MHz. This is, therefore, a new, simple, compact and tuneable source of high purity millimetre-wave signals, capable of providing high power (a high proportion of the power in the unlocked beat signal is transferred to the locked signal).

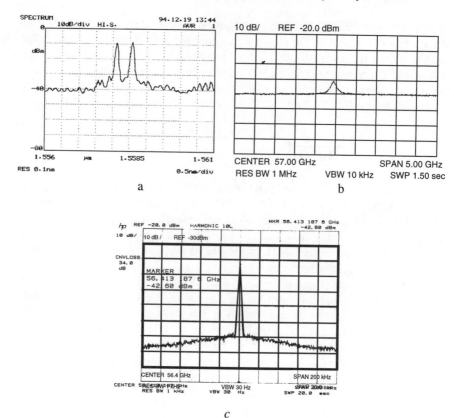

Figure 8.9   *Spectra obtained from dual-mode laser*
(a) Optical spectrum showing mode separation of 0.48nm
(b) Electrical spectrum showing free-running signal at 57GHz
(c) Electrical spectrum showing phase-locked signal at 56.4GHz
using subharmonic electrical injection at 6.3GHz

8.4.5 Edge-entry photodiode

For RF and microwave frequencies the surface-illuminated PIN photodetector offers the best compromise between responsivity and bandwidth [12]. Surface-illuminated detector design requires optimisation of the depletion (absorbing) layer thickness and active area. Thick depletion layers result in high responsivity and low capacitance per unit area but also increase carrier transit times. Optimum performance occurs when the bandwidth due to capacitance and transit time are made equal. Likewise small surface areas result in reduced capacitance but make contacting difficult and ultimately limit the responsivity. Reliability is also an issue since microwave PIN detectors fabricated using mesa etching have exposed p-n junctions. Careful optimisation using planar technology has resulted in reliable PIN detectors with 25GHz bandwidth and a responsivity of 1A/W at a wavelength of 1550nm [13].

The surface-illuminated PIN photodiode can be modified for use at millimetre-wave frequencies by reducing the size of the photosensitive area (for smaller capacitance) and also by thinning the absorber layer (for shorter carrier transit time). The latter modification has the unfortunate side-effect of reducing the sensitivity since less of the incident light is absorbed in a thinner layer. Consequently, the surface-illuminated photodiode suffers from a trade-off between speed and sensitivity. Changing the geometry of the device to edge-entry removes this trade-off. For this geometry, the sensitivity is limited by the coupling efficiency of the optical input into the edge of the device. For relatively low frequencies, this limitation makes edge-entry photodetectors uncompetitive in terms of sensitivity. However, at millimetre-wave frequencies, the edge-entry approach provides higher sensitivity than the surface-illuminated geometry.

In the edge-entry photodiode, light is coupled into an optical waveguide that includes an absorbing layer [14]. In this device, parameters such as the length and width of the waveguide and the absorber layer thickness are carefully optimised to provide the right balance between the various trade-offs involved in the design. Coupling efficiency was enhanced by the insertion of a thick waveguide layer into the structure, analogous to the separate confinement heterostructure (SCH) used successfully in semiconductor lasers for increasing the optical mode size. Bondpad capacitance was reduced by placing the bondpad on a thick dielectric layer, which ensures a total chip capacitance low enough to provide an adequate frequency response when combined with packaging parasitics. A schematic cross-section of this device is shown in Figure 8.10.

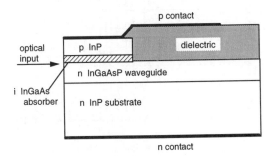

Figure 8.10 *Schematic cross-section of edge-entry photodiode*

Packaging is of critical importance for high frequency operation. The design of the photodetector chip cannot be considered in isolation because of the close interaction of chip parameters with package parasitics. Devices have been mounted in a variety of ways, including packages with coaxial connectors (K or V connectors) and also directly in rectangular metallic waveguide packages - WG22 for Ka band operation (26.5 - 40GHz). A 3dB bandwidth of 50GHz has been achieved in a V-connectorised package at a reverse bias of 1V. Quantum efficiencies of over 50% can be achieved with standard lens-ended fibres.

The dc bias across the photodiode has to be sufficient to provide an electric field for photogenerated carriers to reach the contacts over a time period that is short compared to the period of the modulated optical input signal. Conventional surface-entry photodiodes have relatively thick absorber layers (1 - 5μm) which result in the requirement for bias levels of 5 to 10V. As a consequence of its geometry, the edge-entry photodiode, in contrast, has a relatively thin absorber layer (0.15μm) which allows lower bias levels to be used. A bias of only 1V is sufficient to enable this device to operate at maximum performance, which results in a power consumption reduction of a factor of 5 to 10 over the conventional photodiode. This device will also operate, albeit with slightly reduced performance, without external bias [15]. This is because the built-in potential of the pn-junction is sufficient to provide an adequate electric field on its own to ensure short carrier transit time. In other words, no external bias is necessary, leading to a dc power consumption of zero. This zero bias photodiode was used to show the principle of a radio transmitter with no electrical power consumption, discussed in more detail in the next section. A 3dB bandwidth of around 40GHz has been obtained for zero bias operation. Figure 8.11 shows the variation of electrical output power at 30GHz with optical input power. The line through the low power data points has a slope of 20dB/decade to show any course deviation from linearity.

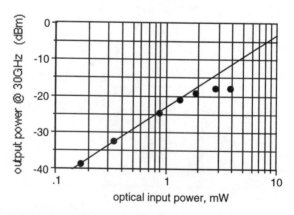

Figure 8.11 *Plot of electrical output power variation with optical input power for zero bias photodiode at 30GHz*

As can be seen from the figure, maximum output power is limited by saturation effects to around -20dBm. However, application of bias removes this saturation and output powers of at least 0dBm can be achieved for a dc power consumption of less than 10mW.

8.4.6 Monolithic optically-preamplified photodiode

The major drawback of the photodetectors described so far is responsivity, since the photodiode is a unity-gain device. Optical preamplification is recognised as an efficient means of improving the signal-to-noise ratio in many systems. A monolithic photodiode and optical preamplifier can be thought of as a photodetector with a high effective responsivity, where the increased responsivity comes from the optical gain in the preamplifier. Such a component has been designed and fabricated [16]. With this device, high frequency operation has been combined with high responsivity (best results obtained for free space optical input are a bandwidth of 33GHz and a responsivity of 89A/W). The combination of high optical gain and millimetre-wave electrical performance means that this device should be ideal for millimetre-wave RoF applications.

The structure is based upon the ridge waveguide laser, with the photodiode absorbing layer being an unpumped extension of the preamplifier gain layer in an edge-entry configuration. This arrangement ensures good optical coupling and low optical feedback at the interface between preamplifier and photodiode. Figure 8.12 shows the longitudinal cross-section of this device in schematic form. The principal design feature is the use of selective epitaxy of p-type ridges to produce localised type conversion in underlying planar n-type material using a patterned dielectric mask. This design technique creates a dopant distribution that ensures low junction capacitance (for high frequency operation) and good electrical isolation.

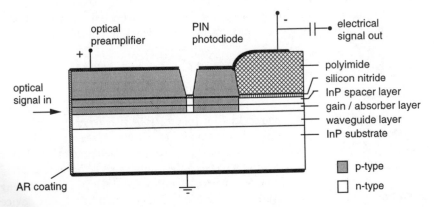

Figure 8.12 *Schematic cross section of monolithic optically-preamplified photodiode*

As in the edge-entry photodiode described above, the bondpad sits on a thick dielectric layer to give low capacitance.   Further design features have been incorporated that give this device useful characteristics such as travelling wave amplification, polarisation insensitivity, wide optical bandwidth and high saturated output power. The best results obtained for devices working at 1550nm are listed in Table 8.1.

Table 8.1   *Results obtained for monolithic optically-preamplified photodiode*

Preamplifier Gain (optical)	18dB
Photodiode Bandwidth	33GHz
Polarisation Sensitivity	1dB
Gain Ripple	1dB
Preamplifier Bandwidth	60nm
Saturated rf output power	0dBm

The major drawback, however, is noise which is due to spontaneous emission from the optical amplifier. In most RoF systems, the characteristics of the radio link (thermal noise in the radio receiver) dominate overall   performance, especially when radio range is pushed towards the limits. In these cases, the high noise levels from the optically-preamplified photodiode will not degrade overall system performance. In other cases, the gain of the preamplifier must be reduced so that the noise level in the device is brought down to a point where it no longer dominates the overall noise perfromance.

Saturation effects in the amplifier and photodiode result in a maximum output power of around 1mW, which is adequate for the short-range applications best suited to millimetre-wave frequencies. The high sensitivity afforded by optical preamplification would find greatest use in passive optical network (PON) architectures, where a high split ratio means that incoming optical signal levels are low.

8.4.7 PhotoHBT

The heterojunction bipolar phototransistor (photoHBT) is another example of a photodetector with gain. In this case the gain is electrical and takes place through transistor action. Conventionally, this device is simply a heterojunction bipolar transistor (HBT) made with a window area in the emitter metal for optical input. The InGaAs layer used in the base-collector of InP-based devices ensures strong absorption at the desired wavelength of around 1550nm.   Although this device can be used simply as a photodetector, it can also be used for applications where advantage can be taken of its dual role as photodetector and active electronic circuit element. An example of this approach is an optically injection-locked oscillator, which is an important component in RoF systems because it allows high-power and low-cost oscillators to be designed without being constrained by purity requirements. In this configuration a locking signal is transmitted over an optical link to transfer the purity of a reference source located centrally in a benign environment to the remote oscillator. Optoelectronic mixers are also important building blocks in many RoF system configurations, where they can be used, for example, to downconvert incoming radio signals at a radio transmitter to IF or baseband to allow a simple low frequency return path. The photoHBT is an ideal device for these applications because it allows good optical coupling from single-mode fibre and can be made with high gain and good frequency response. The third terminal on this type of device is employed to provide access for mixing signals or feedback, and the inherent gain provides high mixing efficiency or the required conditions for oscillation.

Unoptimised three-terminal photoHBTs have been fabricated and assessed for optoelectronic mixing efficiency [17]. Optically and electrically pumped mixing schemes were investigated and results were compared with several types of unipolar (FET) device. The photoHBTs had the better performance as a result of their superior electrical characteristics combined with high-efficiency optical access. Maximum mixing efficiency was obtained at low base-emitter bias, where the base-emitter diode nonlinearity is greatest.

Two-terminal, edge-entry photoHBTs have also been developed in which the base is biased optically [18]. This type of device has high dc gain (greater than 270), and high unity–gain frequency (greater than 30GHz). The high level of performance is due to a combination of having an edge-entry geometry for high efficiency and a two-terminal, optically–biased design for low parasitics. Both of these aspects allow small-area devices to be realised with high efficiency and high speed, whilst keeping fabrication simple. Fabrication is based on the edge-entry photodiode discussed above.

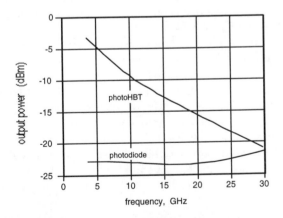

Figure 8.13   *Frequency response of two-terminal, edge-entry photoHBT compared with edge-entry photodiode*

The frequency response for this type of device is shown in Figure 8.13 for an emitter bias of -1V. The optical source had a power of 150μW, a modulation depth around 100% and a wavelength of 1550nm. Also shown in this figure is the response of an edge-entry photodiode for comparison. This photoHBT is attractive as a high-gain photodetector because it gives the same functionality as a photodiode and amplifier combination in a simple, compact and integrated structure.

This device has also been used as part of an indirect optically injection-locked oscillator [19]. The photoHBT provided the locking signal for a 2.1GHz MESFET microstrip oscillator. Locking bandwidths of up to 10% were achieved as a result of the high current gain of the photoHBT. The photoHBT is therefore a potentially useful component for a variety of RoF system configurations. The combination of efficient photodetection, high electrical gain and, in some cases, a third terminal gives this type of device a high degree of performance and flexibility. As a result of this high performance and flexibility, RoF system building blocks such as high-gain photodetectors, optically injection-locked oscillators and optoelectronic mixers have been demonstrated successfully with this type of device.

8.5 System demonstrators

8.5.1 Video transmission at 40GHz

A video transmission experiment was assembled using many of the components and techniques described earlier [20]. This experiment involved the transmission of a video signal superimposed on a millimetre-wave carrier over a length of optical fibre and a radio path. Transmission was performed at 40GHz and used a monolithic optically-preamplified photodiode as the O/E converter. The layout of this system is shown in Figure 8.14.

The optical source used for this demonstration was the FM laser described earlier, which was driven by a microwave oscillator at a frequency of 4GHz. A video signal was applied to the FM input of the oscillator. The isolated output of the laser, which consisted of optical FM sidebands generated by the composite signal, was transmitted along 12.5km of standard single-mode fibre to the photodetector. The dispersion arising from this fibre was sufficient to perturb the phase relationships between individual sidebands so that the output of the photodetector consisted of harmonics of the drive signal with high power levels. In order to account for the multiplication of the composite drive signal, the FM deviation of the input video signal was divided by the number of the selected harmonic.

For the purposes of this demonstration, the radio path was 1 metre long. The radio receiver was tuned to the appropriate harmonic to produce an intermediate frequency suitable for a satellite TV receiver, and the output was displayed on a video monitor. Good quality video transmission was observed, with a comfortable power margin, even though the radio receiver was not optimised for low noise operation. An obvious way to increase the radio range would be to use a power amplifier after the photodetector and/or a low noise amplifier at the radio receiver, since the overall signal-to-noise ratio of the system was not limited by the optical link.

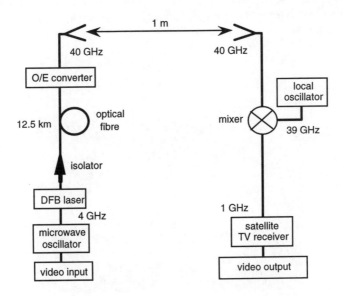

Figure 8.14 *Layout of millimetre-wave RoF video transmission experiment*

8.5.2 Zero electrical power radio transmitter

A second version of the video transmission experiment described above used a zero-bias photodiode as the O/E converter [15]. The transmission frequency was 29GHz (7th harmonic of the drive frequency) to take advantage of compact radio receivers that were available at this frequency. Again, good quality video was observed on the monitor. The potential radio range of this zero-power radio transmitter can be estimated from the saturation characteristics (Figure 8.11), which shows an output power capability of around -20dBm. This power level could give a radio range of several hundred metres, especially if high-gain antennas were used. The important point of this component is its ease of deployment. A radio transmitter with small size, low weight and zero electrical power consumption could be deployed very simply in virtually any location. This would mean that concerns regarding wayleave or power provision when planning installation could be insignificant. Figure 8.15 shows a photograph of the zero electrical power radio transmitter, which illustrates the simplicity of this approach. In this case, the photodiode was mounted in a rectangular metallic waveguide. A single input fibre and a horn antenna output can be seen clearly. No external electrical powering is required.

The potential for low cost is considerable. The chip is simple to manufacture and could therefore be very cheap in quantity. The main cost is associated with the process of optical coupling. Advances in low-cost fibre pigtailing, driven by the requirement for cheap optical sources in an optical access network, should provide a route to bring the cost down enough to enable mass deployment.

Figure 8.15 *Photograph of zero electrical power radio transmitter*

8.6 Conclusion

The potential advantages of using optical links to feed radio transmitters/receivers are becoming clear to both manufacturers and operators, and microwave systems are now being installed. The work described in this chapter addresses some of the problems involved in translating these links to millimetre-wave frequencies, where many of the RoF applications of the future will lie.

In summary:
- only the millimetre-wave frequency bands contain enough available spectrum for the large-scale deployment of radio-based broadband services
- RoF technology can simplify and reduce the cost of remote radio transceivers to an extent where it may be the most cost-effective option for large-scale networks
- very few conventional optoelectronic components are available for operation at millimetre-wave frequencies

- several novel optoelectronic components and techniques have been explored to overcome this limitation
- some of these components and techniques have been used to construct system demonstrators.

It seems inevitable that broadband service delivery to a massive number of mobile or cordless users will one day become a reality. Millimetre-wave radio over fibre technology stands poised to play a crucial role in the infrastructure needed for such a scenario. The novel components and techniques discussed here form a basis for this technology and further development will enable this crucial role to be fulfilled.

8.7   References

1   Cooper, A.J.: 'Fibre/Radio for the provision of cordless/mobile telephony services in the access network', *Electronics Letters*, 1990, **26**, pp2054-2056

2   Georges, J.B., Kiang, M-H, Heppell, K., Sayed, M., and Lau, K.: 'Optical transmission of narrowband millimetre-wave signals by resonant modulation of monolithic semiconductor lasers' *IEEE Photonics Technology Letters*, 1994, **6**, pp.568-570

3   Walker, N.G., Wake, D., and Smith, I.C.: 'Efficient millimetre-wave signal generation through FM-IM conversion in dispersive optical fibre links', *Electronics Letters*, 1992, **28**, pp.2027-2028

4   Scott, D.C., Plant, D.V., and Fetterman, H.R.: '60GHz sources using optically driven heterojunction bipolar transistors', *Applied Physics Letters*, 1992, **61**, pp.1-3

5   Ogawa, H., and Kamiya, Y.: 'Fibre-optic microwave transmission using harmonic laser mixing, optoelectronic mixing, and optically pumped mixing', *IEEE Trans. Microwave Theory and Techniques*, 1991, **39**, pp.2045-2051

6   Sherlock, G., Wickes, H.J., Hunter, C.A., and Walker, N.G.: "High speed, high efficiency, tunable DFB lasers for high density WDM applications", ECOC'92, 1992, paper Tu P1.1

7   O'Reilly, J.J, Lane, P.M., Heidemann, R., and Hofstetter, R.: 'Optical generation of very narrow linewidth millimetre-wave signals', *Electronics Letters*, 1992, **28**, pp.2309-2311

8   O'Reilly, J.J, Lane, P.M., Capstick, M.H., Salgado, H.M., Heidemann, R., Hofstetter, R., and Schmuck, H.: 'RACE R2005: microwave optical duplex antenna link', *IEE Proc. J*, 1993, **140**, pp.385-391

9   Moodie, D.G., Wake, D., Walker, N.G., and Nesset, D.: 'Efficient harmonic generation using an electroabsorption modulator', *Photonics Technology Letters*, 1995, **7**, pp.312-314

10 Lima, C.R., Wake, D., and Davies, P.A.: 'Compact optical millimetre-wave source using a dual-mode semiconductor laser', *Electronics Letters*, 1995, **31**, pp.364-366

11 Willems, J., David, K., Morthier, G., and Baets, R.: 'Filter characteristics of DBR amplifier with index and gain coupling', *Electronics Letters*, 1991, **27**, pp.831-833

12 Bowers, J.E. and Burrus, C.A' Ultra wideband long wavelength PIN photodetectors', *Journal of Lightwave Technology*, 1987, **5**, pp.1339-1350

13 Wake, D., Walling, R.H., Henning, I.D., and Parker, D.G.: 'Planar junction, top-illuminated GaAs/InP PIN photodiode with bandwidth of 25GHz', *Electronics Letters*, 1989, **25**, pp.967-968

14 Wake, D., Spooner, T.P., Perrin, S.D., and Henning, I.D.: '50GHz InGaAs edge-coupled PIN photodetector', *Electronics Letters*, 1991, **27**, pp.1073-1074

15 Wake, D., Walker, N.G., and Smith, I.C.: 'Zero bias edge-coupled InGaAs photodiodes in millimetre-wave RoF systems', *Electronics Letters*, 1993, **29**, pp.1879-1880

16 Wake, D.: 'A 1550nm millimetre-wave photodetector with a bandwidth-efficiency product of 2.4THz', *Journal of Lightwave Technology*, 1992, **10**, pp.908-912

17 Urey, Z., Wake, D., Newson, D.J., and Henning, I.D.: 'Comparison of InGaAs transistors as optoelectronic mixers', *Electronics Letters*, 1993, **29**, pp.1796-1797

18 Wake, D., Newson, D.J., Harlow, M.J., and Henning, I.D.: 'Optically-biased edge-coupled InP/InGaAs heterojunction phototransistors', *Electronics Letters*, 1993, **29**, pp.2217-2219

19 Sommer, D., Gomes, N.J., and Wake, D.: 'Optical injection locking of microstrip MESFET oscillator using heterojunction phototransistors', *Electronics Letters*, 1994, **30**, pp.1097-1098

20 Wake, D., Smith, I.C., Walker, N.G., Henning, I.D., and Carver, R.D.: 'Video transmission over a 40GHz radio-fibre link', *Electronics Letters*, 1992, **28**, pp.2024-2025

Chapter 9

Optical generation and delivery of modulated mm-waves for mobile communications

J.J. O'Reilly, P.M. Lane and M.H. Capstick

9.1 Introduction

There is considerable interest in utilising the relatively uncluttered mm-wave region of the radio spectrum for the delivery of mobile telecommunication services. Service providers are looking at this region due to the congestion at lower frequencies that severely limits the possibility of new allocations and in any case limits the available bandwidth. The limited propagation distances achievable at mm-wave frequencies also offers the advantage of efficient frequency re-use by the adoption of a micro-cellular-based architecture, which reduces the total spectrum required to obtain a given coverage. There is anticipated growth in the requirement for a mobile system that can support tetherless access to broadband fixed network services such as B-ISDN and mm-waves offer the only practicable mechanism for the conveyance of these wide bandwidth signals. Other advantages of the characteristics of a micro-cellular mm-wave system are the limited transmit power that is required, which leads to longer portable battery life, and the small size of antennas which, taken together, can lead to terminal equipment that is smaller, lighter and less obtrusive than current mobile communication equipment. The adoption of a mm-wave based system requires a method whereby modulated mm-wave signals can be generated at remote antenna units. Since, as we have already noted, mm-wave propagation characteristics are severely limited, there will be a need for a large number of these antenna units. Each unit must therefore be of as low a cost as possible. The remainder of this introductory section will outline possible electrical methods for distributed mm-wave generation and briefly outline some of the advantages that could be enjoyed if an optically-based method for the generation and distribution of mm-waves were to be adopted.

9.1.1 Electrical methods for the generation and distribution of mm-waves

The distribution of mm-waves via the use of transmission lines is not feasible due to the very high loss associated with such lines and furthermore the high cost of such cables or waveguides. This renders the central generation and distribution of mm-waves impracticable. Electrical distribution can be achieved by distributing baseband signals or a low intermediate frequency (IF) (whichever is appropriate for the application concerned) from the base unit / switching centre to the individual base stations. There is a wealth of information on the distribution of such signals available from the cable TV industry documenting the distance between repeaters, the maximum number of repeaters and the overall distance before the signal is degraded significantly, as well as economic data. The baseband or IF signals would then have to be upconverted to the required mm-wave frequency at each base station, amplified and radiated. The topology of such a system can be seen in Figure 9.1.

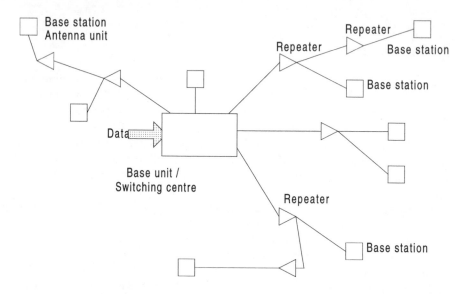

Figure 9.1 *Topology for electrically distributed data for transmission at mm-wave frequencies*

For a typical bandwidth of 0 - 500MHz, such as is used in trunk CATV networks, low-loss cabling is utilised and repeaters must be included at least every 500m to maintain the signals with appropriate signal-to-noise ratios. The complexity and linearity requirements of these repeaters will be very dependent on the nature of the signals; for instance a single channel with a constant envelope modulation scheme or a channel of broadband data will be much simpler to deal with in terms of linearity than multitone frequency division multiplexed (FDM) signals or a single channel where linear modulation of amplitude is utilised. The system design and specification becomes most critical where distortion due to non-linearity can have a potentially catastrophic effect on signal-to-noise and distortion (SINAD) ratios. Whatever the signal type the repeaters must be equipped with equalisers to counteract the frequency response of the cables used, as well as the power amplification stages. Inevitably there will be a loss of fidelity in the repeating process with the signals never being restored to their original quality. This therefore imposes an upper limit on the number of stages that can be realistically cascaded and thus the area over which a single base unit and switching centre can operate. This can be overcome when a single digital data channel is being used by regenerating the data every so often such as is the practice in digital telephony circuits. Additionally there are questions of electromagnetic compatibility with the potential of interference to and by the signals in the cables. Cables also have the problem that if any moisture is allowed to ingress then performance degrades very rapidly, the only solution being to replace the length of cable in question.

Once distribution has been achieved the base stations translate the base band or IF signals up to the required mm-wave frequency. The required circuit blocks for a remote base station can be seen in Figure 9.2.

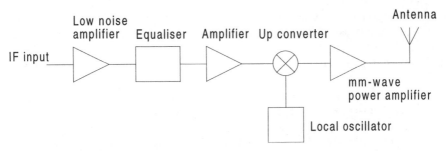

Figure 9.2 *Base station topology*

On the whole the topology employed is independent of the number of channels and modulation scheme(s) utilised. However, these details will greatly influence the specification required of the various circuit blocks. In all cases a low-noise amplifier with good dynamic range will be required and an equaliser much the same as in the repeaters for the truck distribution network. These equalised signals will then have the intended spectral density and relative amplitudes for transmission over the free space link. Signals are then amplified up to the maximum level possible that would allow maintenance of performance criteria of the upconversion mixer. The philosophy for having as much gain as possible at the low IF frequency is the ready availability and low cost of amplifiers at these frequencies in comparison to those at the mm-wave output frequency. The prime mixer consideration would have to be linearity particularly when dealing with FDM signals. The mm-wave amplifier will similarly have linearity requirements governed by a combination of the number of channels and modulation formats used. This then leads to the requirement for a local oscillator, which has the potential to be the most complex and expensive circuit block. Again the complexity can to some extent be related to the nature of the signals to be radiated; if signals are truly broadband then the requirements in terms of stability, phase noise and frequency accuracy of the local oscillator will not be as stringent as compared to applications with many narrower band signals. For broadband applications it may be possible to utilise free running oscillators such as dielectric resonator oscillators (DRO) but these rarely achieve better than 0.1% overall frequency accuracy, so the modulation bandwidth must be large in comparison to this for such a system to be applicable. For more usual signal bandwidths, good stability and accuracy must be achieved and are generally only available from oscillators that are locked to some lower frequency standard; by locked, here, we mean that it will almost certainly have to be phase- rather than frequency-locked to achieve the phase noise required for narrower band signal applications.

Summarising, when opting to use electrical distribution of signals there would be a requirement for frequent signal amplification and equalisation to maintain usable signal amplitudes especially at the higher frequency end of the bands used. The base station equipment would be complex with tight performance requirements. This therefore leads to a system that devolves a lot of the complexity to each base station and a system where the signal quality degrades relatively rapidly as the distance from the switching centre increases, limiting the area that can be controlled from one switching centre. Finally the coaxial cable used has the potential to suffer from or create electromagnetic compatibility problems.

As was demonstrated above, electrical methods for the remote generation of modulated mm-waves are severely limited in both performance and cost, which prompts a search for other alternative techniques. An obvious candidate is the adoption of a fibre-based method. Optical fibre offers very low loss (about 0.2 dB/km) and a bandwidth that is measured in THz. An optical network could convey appropriate signals to a simple antenna unit where the mm-wave signals could be directly generated. The cost of these antenna units can be low as soon as optoelectronic integrated circuit (OEIC) technology matures, with packaging and fibre alignment being key considerations in this respect. Erbium-doped fibre amplifiers (EDFAs) are now a mature technology and their use offers the possibility of dividing the optical signal to many antenna units. This is one of the most exciting possibilities associated with the adoption of a fibre-based solution. The sharing of a centralised generation system between many antenna units offers the possibility of very low system cost as well as excellent performance due to both the concentration of resource and the possibility of tight environmental control. In the next section the methods available for the optical generation of modulated mm-waves will be described.

9.2 Optical mm-wave generation methods

In this section we will first discuss the methods that are available for the generation of unmodulated mm-waves and then go on to discuss the imposition of modulation. Some of the impairments associated with fibre-based systems will then be discussed and the section will conclude with a summary of the methods described. It is convenient to divide the methods into those that are based on a single laser, those that are based on two lasers and those that are based on frequency multiplication.

9.2.1 Single laser methods

The simplest method for the optical generation of mm-waves is to directly modulate the intensity of the output of a laser. This can be achieved in two ways; the laser bias current can be modulated or the laser can be operated in cw mode and an external modulator can be used to modulate the intensity of the resulting output as shown in Figure 9.3.

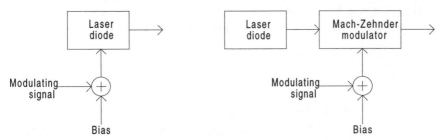

Figure 9.3 *Direct modulation of the optical intensity*

For mm-wave generation the modulating signal would have to be at a mm-wave frequency and bear information as required. After transmission through a fibre and direct detection on a PIN photodiode the photocurrent would be a replica of the modulating signal applied either directly to the laser or to the external modulator. This photocurrent would undergo transimpedance amplification to yield a voltage that would be used to excite an antenna. The direct modulation of the laser bias level is only suited to operation out to high microwave frequencies with the fastest lasers currently reported having a modulation bandwidth of about 20 GHz [1]. The non-linearity of the laser, which leads to intermodulation products, also makes this method inherently noisy as soon as the drive signal consists of modulated carriers. Only a few experiments, all at relatively low frequencies, have been reported for this method; Cooper, for example, described a CT-2 compatible link operating at 864-868 MHz [2]. Mach-Zehnder modulators, on the other hand, have been reported with bandwidths that are compatible with the requirements of mm-wave systems. For example, Walker [3] has described a device with a 36 GHz modulation bandwidth and Noguchi *et al.* [4] reported a device with a 40 GHz bandwidth and a half-wave voltage of 3.6 V. A novel antenna-driven modulator has been reported by Sheehy *et al.* [5] that had a bandwidth of 94 GHz and the authors know of research travelling-wave devices with bandwidths approaching 100 GHz. A drawback of Mach-Zehnder modulators is the increasing drive voltage requirements of higher bandwidth modulators. This has serious implications for the complexity and hence the cost of the drive amplifier. For cw applications the drive amplifier can be relatively straightforward, based, for example, on a class C amplifier. However, for the generation of modulated mm-waves the drive amplifier must compensate for the inherent static and / or dynamic non-linearity of the Mach-Zehnder modulator. This suggests a very complex drive arrangement which will be discussed in more detail in the section addressing the imposition of modulation. To our knowledge no system experiments assessing the suitability of Mach-Zehnder-based systems for the direct generation of modulated mm-waves have been reported.

The two methods described above relied on the direct detection of the intensity of the optical field to recover the mm-wave signal. All but one of the methods that follow rely on the principle of coherent mixing. Accordingly we will digress at this point and describe the general concept of coherent mixing. Two optical fields of angular frequencies Ω_1 and Ω_2 can be written as

$$E_1 = E_{01} \cos(\Omega_1 t)$$
$$E_2 = E_{02} \cos(\Omega_2 t)$$

(9.1)

Adding these fields and detecting the resulting signal on the surface of a PIN photodiode yields a photocurrent that is proportional to the square of the sum of the fields. The only term of interest in the resulting photocurrent is $E_{01}E_{02} \cos((\Omega_1 - \Omega_2)t)$. By controlling the difference in frequency between the two optical fields mm-waves at any frequency can be generated up to the frequency limit of the PIN photodiode.

The first method to use coherent mixing is shown in Figure 9.4.

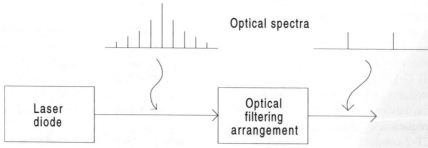

Figure 9.4 *Selection of appropriate FM sidebands of a laser*

The laser is frequency-modulated which leads to an optical spectrum as shown in the figure. A technique that can select two of these modulation sidebands, separated by the required mm-wave frequency, for transmission to the antenna is adopted. At the antenna unit the two signals mix coherently to produce the required mm-wave signal. There are a range of methods that can be used to separate the two signals. Figure 9.5 illustrates these; one is based on the use of a simple optical filter, another uses a semiconductor optical amplifier with an appropriate free spectral range (FSR) and the third uses two mode-locked lasers to generate the two required optical fields. Goldberg [6,7] has investigated these techniques and demonstrated the generation of signals at up to 35 GHz with an electrical linewidth of less than 10 Hz.

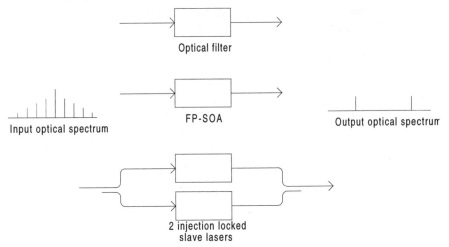

Figure 9.5 *Methods for the selection of the required sidebands*

The final single-laser solution uses the dispersive characteristic of optical fibre to effect a frequency modulation to intensity modulation (FM-IM) conversion [8]. A laser is frequency-modulated, as in the previous method, to yield a set of modulation sidebands. Instead of filtering out two components, all of these modulation sidebands are transmitted through a dispersive length of fibre which alters the relative phases of the components. Mixing of these components on a PIN photodiode yields an intensity-modulated signal at a harmonic of the drive signal applied to the laser. Achieving a reasonable modulation depth at a particular wanted frequency is dependent on the fibre length and the modulation index of the laser. This technique was demonstrated by the transmission of video signals on a 40 GHz carrier [9]. A DFB laser was frequency-modulated with a 4 GHz subcarrier that was itself frequency-modulated by a video signal. A 12.5 km length of fibre provided the necessary dispersion before the signal was detected and radiated by a horn antenna - the antenna's selectivity providing filtering to remove unwanted harmonics. The signal was received after a propagation distance of 1 m, downconverted to a 1 GHz IF and viewed.

9.2.2 Dual laser methods

There are two techniques used to control the relative frequency of a pair of laser diodes to maintain the required frequency offset between two electric fields. These two fields can mix on a photodetector to yield an electrical signal as described above. One method, the optical frequency-locked loop (OFLL), only maintains the correct mean frequency offset and ignores small-scale frequency perturbations caused by phase noise. The other method, the optical phase-locked loop (OPLL), tracks these small scale perturbations. Both of these methods can be represented by the arrangement shown in Figure 9.6.

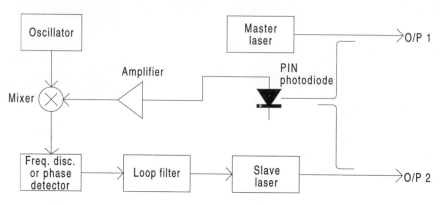

Figure 9.6 *Optical frequency-/phase-locked loop*

Since the instantaneous frequency of the electrical signal generated upon mixing the two optical fields is equal to the instantaneous difference in frequency between these fields, the OFLL, by only maintaining a mean offset, produces an electrical signal with linewidth that is approximately the sum of the linewidths of the individual lasers. On the other hand, the OPLL maintains the required instantaneous frequency offset and is therefore capable of achieving very narrow linewidth electrical signals. This superior performance of the OPLL is obtained at the cost of laser complexity; while OFLLs have been realised with standard, fairly inexpensive DFBs all OPLLs reported to date have required far more complex laser structures such as 3-contact DFBs. This requirement exists because the slave laser in an OPLL must be capable of being tuned at a sufficiently high rate to track the frequency perturbations of the master laser. The wide electrical linewidth of the OFLL can only be alleviated by resorting to very narrow linewidth lasers such as Nd:YAG or erbium ring devices. However, as the source linewidth is reduced, the maximum power that can be conveyed in the fibre without experiencing severe attenuation due to simulated Brillouin scattering (SBS) falls. As an illustrative example, an optical source with a 10 kHz linewidth and an optical power of 10 dBm would experience a power loss of about 7.5dB through SBS-induced back-scattering after propagating through 20 km of fibre. In order not to reach the SBS threshold the power would have to be limited to below about 1.5 dBm. These limitations on power have a significant impact on the deployment flexibility of systems requiring narrow linewidth lasers.

Both Tun *et al.* [10] and Kawanishi *et al.* [11] have reported the use of OFLLs for optical receiver calibration. The reported systems allowed the generation of beat terms at up to 100 GHz. Williams *et al.* [12] and Gliese *et al.* [13] have reported an OPLL that generated electrical signals at up to 30 GHz with mHz linewidths.

9.2.3 Frequency multiplication methods

Two methods will be discussed: both exploit the E-field response of a Mach-Zehnder modulator. The output field of such a modulator can be described by:

$$E_{out}(t) = E_{in}(t)\cos\left(\frac{\pi}{2}\frac{V_{mod}(t)}{V_\pi}\right) \tag{9.2}$$

where $E_{in}(t)$ is the incident optical field applied to the modulator, $V_{mod}(t)$ is the modulating voltage applied to the modulator and V_π is the modulating voltage required for the output to be, in the ideal case, totally suppressed. If a modulating voltage of the form

$$V_{mod}(t) = V_\pi(1+\varepsilon) + \alpha V_\pi \cos(\omega t) \tag{9.3}$$

is applied to the modulator, where ε and α are respectively normalised bias and drive levels, then the output field is:

$$E_{out}(t) = \cos\left(\frac{\pi}{2}[(1+\varepsilon) + \alpha\cos(\omega t)]\right)\cos(\Omega t) \tag{9.4}$$

where Ω is the angular frequency of the applied optical field. This last expression can be expanded as a series of Bessel functions:

$$
\begin{aligned}
E_{out}(t) \;=\; & \frac{1}{2}J_0\!\left(\alpha\frac{\pi}{2}\right)\cos\!\left(\frac{\pi}{2}(1+\varepsilon)\right)\cos(\Omega t) \\
& - \frac{1}{2}J_1\!\left(\alpha\frac{\pi}{2}\right)\sin\!\left(\frac{\pi}{2}(1+\varepsilon)\right)\cos(\Omega t\pm\omega t) \\
& + \frac{1}{2}J_2\!\left(\alpha\frac{\pi}{2}\right)\cos\!\left(\frac{\pi}{2}(1+\varepsilon)\right)\cos(\Omega t\pm 2\omega t) \\
& - \frac{1}{2}J_3\!\left(\alpha\frac{\pi}{2}\right)\sin\!\left(\frac{\pi}{2}(1+\varepsilon)\right)\cos(\Omega t\pm 3\omega t) \\
& + \;\cdots
\end{aligned}
\tag{9.5}
$$

which gives the levels of the optical spectral components for any arbitrary drive and bias. There are two cases of particular interest - these are at $\varepsilon = 1$ and at $\varepsilon = 0$ which give rise to the frequency multiplicative methods that are referred to as the 2-f and 4-f methods respectively.

9.2.3.1 The 2-f method
The arrangement used to realise this method is shown in Figure 9.7 and comprises a laser operated in cw mode and a Mach-Zehnder modulator with drive amplifier.

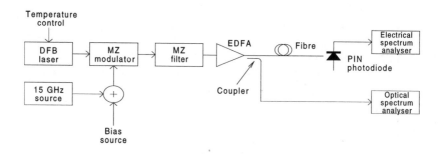

Figure 9.7 *Topology of the 2-f and 4-f generation methods*

It can be seen from the equation describing the output spectrum that if the modulator is biased at $\varepsilon = 1$ then the central component at Ω is suppressed as are all the other even terms. The field is dominated by two components at $\Omega \pm \omega$ which when mixed on a PIN photodiode lead to an electrical signal at 2ω. This doubling in the frequency of the generated signal compared to the drive applied to the modulator leads to the 2-f term used to describe this technique. It should be emphasised at this point that the two optical components generated using this method are derived from the same optical source and therefore have identical phase noise. Upon mixing on the photodetector this yields an electrical signal with a linewidth dependent only on the signal purity of the source used to drive the Mach-Zehnder modulator. Of course there are other components in the spectrum but these are at a much lower level than the two main wanted components.

Figure 9.8 shows the electrical spectrum at the output of a PIN photodiode when the modulator was driven at 15 GHz; it can be seen that the 30 GHz signal has a very narrow linewidth limited only by the resolution bandwidth of the spectrum analyser.

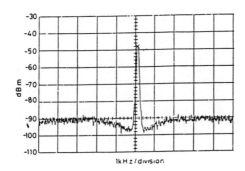

Figure 9.8 *Electrical spectrum at the PIN photodiode for the 2-f method (15 GHz drive)*

9.2.3.2 The 4-f method

The arrangement used to demonstrate this method is identical to that used for the 2-f method except that the drive and bias levels are different. If the modulator is biased at $\varepsilon = 0$ then the odd components are suppressed. The optical spectrum consists of a central component and two other components separated by 2ω either side of this central component. By adjusting the drive level appropriately the central component can be suppressed. This occurs when $J_0(\alpha\pi/2)$ is zero which corresponds to a drive level, α, of about 1.53. The resulting optical spectrum now consists of two components separated by four times the drive frequency of the modulator which yield an electrical signal at this difference frequency after mixing on a PIN photodiode. Again the phase noise on these components is totally correlated.

Figure 9.9 shows the optical spectrum at the PIN photodiode for the 4-f method and Figure 9.10 shows the corresponding electrical spectrum. A very narrow linewidth 60 GHz signal is generated for a 15 GHz drive signal.

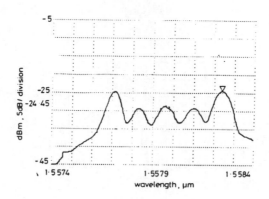

Figure 9.9 *4-f optical spectrum (15 GHz drive)*

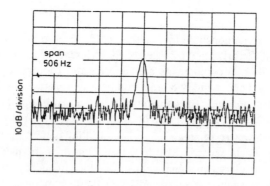

Figure 9.10 *60 GHz electrical spectrum*

We will now describe how modulation can be impressed such that it appears on the electrical signal generated using the above methods.

9.2.4 Modulation

An important aspect of the remote delivery of mm-waves using optical fibres is that the mm-waves generated at the remote site should be modulated with the message information, otherwise all that has been achieved is the remote delivery of a carrier, leaving the data to be delivered and modulated on to the carrier by some other means. This section addresses the various options available for the modulation of the optical signals such that the mm-wave generated at the remote site will embody the message information. The main engineering issues will be discussed using four possible cases, single carrier (with data) and multiple carriers for both direct mm-wave modulation and modulation at a lower frequency. To begin to address this question we must first consider the system topologies for direct modulation and low-frequency modulation of the optical carrier, in terms of the required circuit blocks. We hope to show that the choice of topology can be condensed into a few simple criteria, namely availability and specification of optical modulators and RF / mm-wave amplifiers.

9.2.4.1 Direct modulation
A typical system utilising direct modulation of a single optical source can be seen in Figure 9.11.

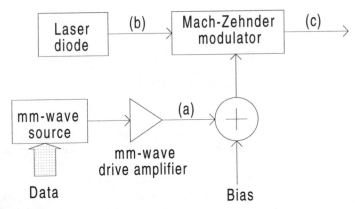

Figure 9.11 *Direct modulation*

To appreciate the hardware implications of this choice of topology firstly consider the signal spectra involved. Figure 9.12 shows the spectra at the points labelled in figure 9.11; ω_{mm} is the mm-wave frequency and Ω is the frequency of the single optical carrier. There are modulation sidebands shown on the mm-wave signal; these can be taken to mean either a single channel or multiple channels of data. For a direct modulation scheme the optical signal is directly detected to generate the mm-wave at the remote site.

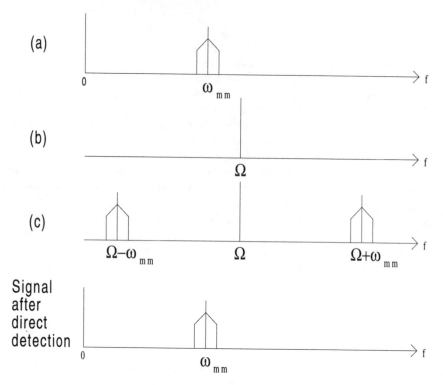

Figure 9.12 *Signal spectra*

For direct modulation of the optical carrier the requirement is for a Mach-Zehnder modulator with a bandwidth in excess of the required mm-wave frequency or a modulator specially designed to have good performance over a range of frequencies centred on the required mm-wave frequency. Also a mm-wave power amplifier will be required, capable of driving the modulator to the required depth of modulation. The mm-wave carrier must be modulated with the data channel(s) so that the resultant optical modulation will contain the data to be distributed over the mm-wave link. The linearity requirements of both the optical modulator and drive amplifier are dependent on the nature of the signals to be distributed and could be technically demanding and difficult to achieve. At the far end of the link the optical receiver simply uses direct detection of the amplitude-modulated optical signal to reproduce the mm-wave signal; the amplitude of this mm-wave is proportional to the power in the optical modulation sidebands and the responsivity of the photodiode.

9.2.4.2 Low-frequency modulation

Low-frequency modulation is the term we have chosen to describe the modulation of a dual-frequency optical source; the topology can be seen in Figure 9.13.

The system using low-frequency modulation shown in Figure 9.13 utilises a dual-frequency optical source which could be realised using any of the methods outlined in section 9.3; the modulation is then applied to one of the optical carriers. The signal spectra resulting from this system are outlined in Figure 9.14. where Ω_1 and Ω_2 are the optical frequencies and ω_{lf} is the modulating signal. The spectra relate to the points indicated on Figure 9.13 with the addition of the spectrum of the signal detected at the remote base station, this signal being the result of coherently detecting the optical signals, which results in convolution of the signals.

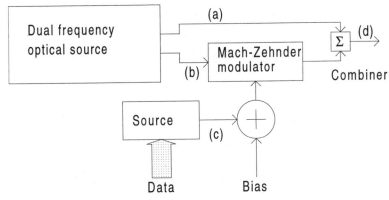

Figure 9.13 *Low frequency modulation*

The advantages of this approach for modulation of the optical signal are many, though it will result in the trade-off of additional complexity in the optical source. The salient points are that a low-frequency optical modulator will usually have a lower V_π and therefore a lower drive-level requirement, the modulator will be more readily linearisable using the available techniques and linear drive amplifiers of high intercept points are more readily available. The frequency chosen for this modulation is constrained in the main by the performance available from the mm-wave filters, as the method produces two sidebands with data and possibly a carrier component at $\Omega_2 - \Omega_1$ (depending on the bias point chosen for the modulator). This double sideband nature of optical amplitude modulation requires that one of the sidebands be rejected. This usually results in a choice of frequency in the low GHz region, which would allow low-loss bandpass filtering. The choice of too low a frequency would result in filters with narrow bandwidth and therefore very high insertion loss specifications. There is in fact no requirement for filtering at the mm-wave frequency if a single channel is used and the modulation is at base band; this, however, would not be the case in the majority of the envisaged applications.

At the far end of the link the optical receiver employs optical mixing, equivalent to coherent detection, to generate the mm-wave signal; the amplitude of the signal is then proportional to the electrical field strength in both the components to be mixed along with the responsivity of the photodiode.

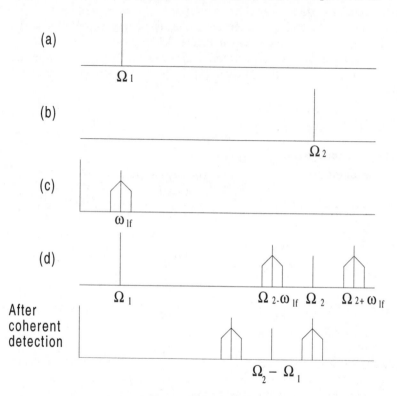

Figure 9.14 *Spectra resulting from the low frequency modulation of a dual frequency optical source*

9.2.4.3 Comparison of the modulation methods
When seeking the best option for a particular application all the points outlined above must be considered and additionally the issue of the extra complexity required to produce a dual-frequency optical source. It is worth noting that this sub-system would only be required once for every switching centre so when compared to the number of base stations that could be supplied the additional complexity may well have a minor impact on the overall technical difficulty.

Table 9.1 provides a summary of the issues considered to be of greatest importance.

Table 9.1 *Factors important in the choice of modulation scheme*

	Single channel	Multi channel
Direct modulation	**Constant envelope modulation:** Modulated mm-wave source, mm-wave power amplifier, Optical modulator with BW > mm-wave frequency required. **Analogue modulation:** Modulated mm-wave source, Linear mm-wave power amplifier, Linearised optical modulator with BW > mm-wave frequency (Linearised for optical power)	Generation of multi-channel mm-wave signals. Linear mm-wave power amplifier. Linearised optical modulator with bandwidth in excess of the required mm-wave frequency, (linearised for optical power)
LF modulation	**Constant envelope modulation:** Drive amplifier with bandwidth suitable for the IF or base-band signals (UHF or low microwave). Low-frequency optical modulator. Drive amplifier and optical modulator at half the mm-wave frequency. **Analogue modulation:** Linear drive amplifier. Linearised low frequency optical modulator (E-field linearised). Drive amplifier and optical modulator at half the mm-wave frequency.	Linear power amplifier (UHF or low microwaves). Linearised low-frequency optical modulator (E-field linearised). Drive amplifier and optical modulator at half the mm-wave frequency.

In essence we can conclude: if multiple channels are to be transmitted, then low-frequency modulation becomes the strongest candidate especially when considering the price and availability of linear mm-wave amplifiers and optical modulators.

9.2.4.4 Modulation formats

The modulation formats that can be used on the carrier or subcarriers are only really limited by the linearity of the system, and poor linearity is only allowable in the situation where a single channel of data is to be used. If a single carrier is employed and the modulator drive amplifier is used in saturation, along with an unlinearised optical modulator, then the modulation format must be limited to one utilising a constant envelope. For the case where the modulation is linear both systems of modulation (direct and low frequency) will be transparent to the modulation imposed on the carrier or sub-carriers. There may be potential problems with some modulation schemes if the phase noise of the optical source is excessive; if this is the case, detailed consideration must be given as to whether or not this will impact on the overall performance of the system.

9.2.5 Impact of differential delay

Up to this point we have assumed that signals that are generated with phase noise coherence will interact on the PIN photodiode with the same degree of coherence. In reality this is not the case and a differential delay, Δt, between the two components will cause a degradation in the electrical linewidth as Figure 9.15 indicates.

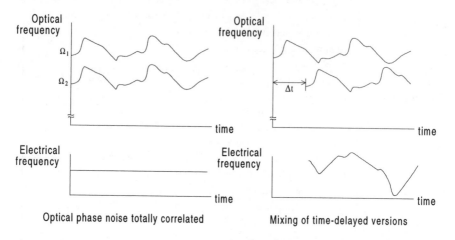

Figure 9.15 *Illustration of differential delay leading to phase noise decorrelation*

Differential delay can be incurred during the modulation process due to the physical path length difference experienced by the two optical components. Fibre dispersion will also cause differential delay due to the different propagation speeds of the two optical spectral components. The problem of the linewidth of the signal resulting from the mixing of two signals has been analysed by Gallion and Debarge [14]. Here we will only summarise the final result of their work which is an expression for the spectrum, $S_I(\overline{\omega})$, of the detected photocurrent:

$$S_I(\overline{\omega}) = (1+\alpha^2)^2 \, \delta(\overline{\omega}) + \alpha^2 \exp(-\overline{\tau}_0)\delta(\overline{\omega}-\overline{\Omega})$$

$$+\alpha^2 \exp(-\overline{\tau}_0)\frac{1/\pi}{1+(\overline{\omega}-\overline{\Omega})^2} \cdot \left[\exp(-\overline{\tau}_0) - \frac{\sin\big((\overline{\omega}-\overline{\Omega})\overline{\tau}_0\big)}{\overline{\omega}-\overline{\Omega}} - \cos\big((\overline{\omega}-\overline{\Omega})\overline{\tau}_0\big) \right]$$

$$(9.6)$$

where α is the amplitude ratio between the two fields, $\overline{\omega}$ is the normalised electrical frequency ($\overline{\omega} = \omega/2\gamma$) where γ is the full angular linewidth at half maximum (FWHM) of the laser, $\overline{\Omega}$ is the normalised separation in frequency between the two lasers ($\overline{\Omega} = \Omega/2\gamma$) where Ω is the absolute frequency separation of the two lasers. $\overline{\tau}_0$ is the normalised time delay between the two signals ($\overline{\tau}_0 = 2\gamma\tau$) where τ is the absolute time delay. The first term in this expression is a dc term which will be ignored hereafter; the second term represents a monochromatic component and the third term is an approximately Lorentzian term. As decorrelation increases there is an uniform transfer of energy from the monochromatic term to the Lorentzian term. Figure 9.16 plots the Lorentzian part of the spectrum for values of $\overline{\tau}_0$ of 1, 2, 4 and 8 against normalised offset frequency.

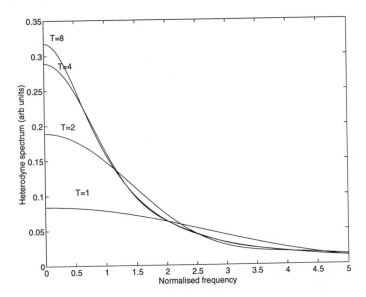

Figure 9.16 *Spectrum of the Lorentzian component of the electrical spectrum*

As can be seen from the figure the linewidth of this component decreases as the decorrelation increases but the level increases. To illustrate this effect we plot in Figure 9.17 the electrical spectra for lasers of 1 MHz linewidth and 10 MHz linewidth for differential delays of 5 ns and 50 ns corresponding to approximately 1 m and 10 m of differential propagation distance in fibre. As can be seen the electrical spectrum generated by the narrower linewidth laser is almost unaffected by the delay and shows a very slight increase in noise floor at the longer delay while the wider linewidth laser system suffers a severe performance penalty at this delay. It is worth noting that the wider linewidth laser performs very well at the lower delay.

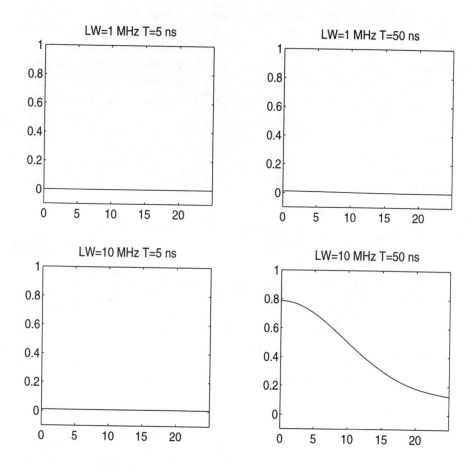

Figure 9.17 *Electrical spectra for coherent systems with 1 m and 10 m of differential delay. Plots show spectrum relative to monochromatic component against frequency in MHz.*

Fibre dispersion can also cause differential delay. For standard monomode fibre with a dispersion parameter of 15 ps/(nm.km) for a system operating at 1.55 μm, components separated by 60 GHz would experience a delay of about 11 ps/km of propagation. This phenomenon is therefore of much lower significance than absolute differential delays but does provide a fundamental limitation to the performance of optical mm-wave generation methods.

9.2.6 Summary of methods for optical generation of mm-waves

Table 9.2 *Comparison of optical mm-wave generation methods*

Method	Frequency	Complexity	Imposition of modulation	Linewidth	Notes
Direct laser mod.	< 20 GHz	Low	Easy	Poor	1
External laser mod.	< 60 GHz	Medium	Easy	Good	2
FM and select sidebands	Limited by PIN ?	High	Medium	Excellent	3
FM-IM	Limited by PIN ?	Medium	Easy but limited (FM only)	?	4
OFLL	Limited by PIN	Medium	Easy	See note 5	
OPLL	Limited by PIN	High, see note 6	Easy	Excellent	
2-f	Limited by PIN	Medium	Medium, see note 7	Excellent	
4-f	Limited by PIN	Medium	Medium, see note 7	Excellent	

Notes
1. Unsuited to almost all applications due to the very noisy electrical signals caused by the laser nonlinearity. Cannot really support multi-channel operation.
2. Achieving a well defined nonlinearity in a high power, mm-wave drive amplifier is very difficult. It appears likely the Mach-Zehnder modulator bandwidths will be improved to the point where the PIN photodiode is the upper frequency limiting factor.
3. Method is really of academic interest - too complex and unstable for field use.
4. Can only support FM signals. Questionable if it can support multi-channel operation. Needs a tuned length of fibre therefore cannot form the basis of a distribution network serving many antennas at different distances from the base unit.
5. Can achieve very good electrical linewidths with very high quality lasers. However with these very narrow linewidth lasers stimulated Brillouin

scattering will severely limit the optical power that can be used and hence the splitting ratio achievable for serving multiple antennas.

6. OPLLs require exotic lasers. At the moment these are only available as research devices.

7. The 2-f and 4-f methods differ from the OFLL/OPLL in that in general the two optical signals need to be separated by optical filtering before modulation can be applied. This implies a need for a stabilisation circuit to maintain the correct laser frequency relative to the optical filter.

9.3 MODAL - Microwave Optical Duplex Antenna Link

MODAL is a RACE II (Research and Development for Advanced Communications in Europe) funded research project that is investigating the application of optical mm-wave generation techniques to the provision of mobile telecommunication services in Europe. The overall objective of the RACE programme is:

> *Introduction of Integrated Broadband Communication (IBC) taking into account the evolving ISDN and national introduction strategies, progressing towards Community-wide services by 1995.*

The definition of *integrated* is given by RACE as proper interworking of all the constituent parts of the complete European network including mobile whilst *broadband* is defined as the delivery of over 140 Mbit/s at the customer interface. For mobile communications it is expected that the Universal Mobile Telecommunication System (UMTS) will provide most of the required capacity. UMTS, however, is expected to operate at low microwave frequencies and cannot, due to bandwidth limitations, support broadband. A mm-wave system, Mobile Broadband System (MBS), is suggested to deliver broadband services and there is also a potential need for high data-rate broadcast systems for applications as diverse as the provision of entertainment services and the operation of traffic telematic systems.

The MODAL project aims to contribute to the development of these systems by investigating optical fibre-based options for the connection of remote antenna units to a central base unit. MODAL takes this further by also investigating optical methods for the generation of the required mm-wave signals. The original MODAL consortium comprised: Alcatel-SEL, Stuttgart, Germany; University of Wales, Bangor, UK; University of Aveiro, Portugal; National Technical University of Athens, Greece; and CET, Aveiro, Portugal. An extension of the MODAL project involved the addition of new members to the consortium in 1994. These new members are: GEC-Marconi Materials Technology, Caswell, UK; Thomson-CSF, France; Lille University, France; and The Fraunhoffer Institute, Freiburg, Germany.

The original MODAL project was to demonstrate a duplex fibre-supported link operating at 30 GHz. The decision to operate at 30 GHz was taken due to the availability of commercial opto-electronic devices for this frequency range. The extended project increased the frequency of operation to 60 GHz and also included some opto-electronic device and MMIC development to support operation at this higher frequency. MODAL is seen as having a wide range of possible application areas including, for example, mobile communications, tetherless access to the IBC network, the provision of microcells in high traffic density sites, entertainment broadcast and traffic telematic systems.

9.3.1 System topology

For MODAL a forward link system topology based around a dual-frequency optical source and a linear low-frequency optical modulator was selected. This approach was chosen as the most feasible for use with modulated sub-carriers, as the requirements on linearity would be difficult to achieve by other methods requiring higher operating frequencies. This was implemented using an optical wavelength of 1550nm to allow the integration of erbium-doped fibre optical amplifiers (EDFA) into the system to achieve optical signal powers of large enough magnitudes to produce electrical signal powers of a reasonable amplitude when coherently detected. For the reverse link an optical wavelength of 1300nm was chosen, allowing the same fibre to be used for both directions. This reverse link utilises a directly modulated, high quality linear laser diode to minimise intermodulation distortion together with a tuned receiver with low noise and high dynamic range to accommodate a wide range of signal levels associated with signals in the mobile environment. Figure 9.18 shows the basic system configuration.

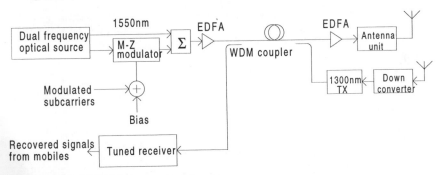

Figure 9.18 *MODAL system topology*

9.3.2 Dual frequency optical source

Several alternative approaches to the realisation of the dual-frequency optical source, outlined above, have been considered as a part of the MODAL project.

The original proposal was to construct an OFLL using two erbium ring lasers (ERLs) that would be fabricated in the consortium [15]. ERLs were made that had a linewidth of about 20 kHz but stability problems were encountered in seeking to maintain the required frequency offset. As has been noted earlier, the use of very narrow linewidth sources such as these ERLs could also lead to a penalty due to SBS. Another alternative considered was the use of very narrow linewidth DFB lasers. These were available in the consortium with a linewidth of about 250 kHz which would have generated a mm-wave signal with a linewidth acceptable for broadband transmission. However, these were research devices and it is unlikely that commercial devices would be available for a deployable system within the 1995 target of RACE. Before the DFB OFLL was implemented the 2-f method was developed and adopted for the MODAL demonstrators. This new technique allowed the use of opto-electronic components that are available now, resolves the SBS problem by removing the need for narrow linewidth sources and greatly simplifies the imposition of linear modulation by only requiring linearity in the relatively low-frequency circuitry that impresses the modulation onto one of the optical components.

9.3.3 The MODAL demonstrators

The MODAL project is developing a family of demonstrators. These show a development of capability that is illustrated in Figure 9.19.

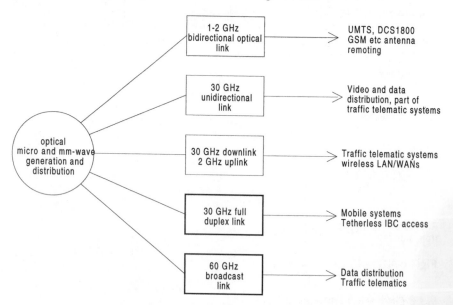

Figure 9.19 *Development of the MODAL demonstrator family*

An asymmetrical system offering high broadcast data-rates at mm-wave frequencies, with a much lower rate channel from the mobile to the fixed system to request data for broadcast, has many applications, particularly in the field of traffic telematic systems. The two major demonstrators in the project are a 30 GHz full duplex system and a 60 GHz broadcast system.

The detailed topology of the 30 GHz full duplex demonstrator is shown in Figure 9.20.

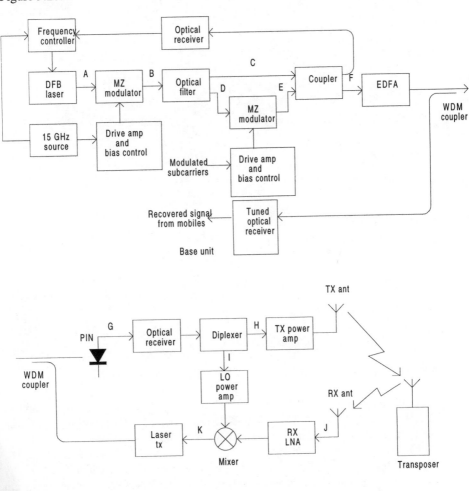

Figure 9.20 *Topology of the 30 GHz bi-directional MODAL demonstrator*

Figure 9.21 shows the expected spectra at the points labelled with upper-case letters in Figure 9.20.

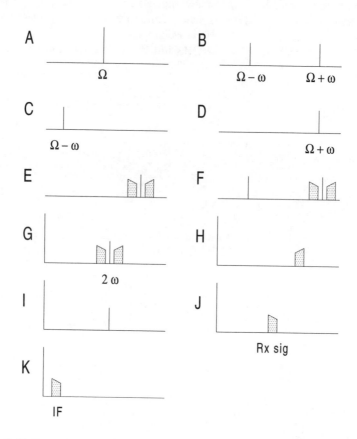

Figure 9.21 *Spectra for the bi-directional MODAL demonstrator*

In the base unit the DFB laser is temperature-controlled to maintain its optical frequency, Ω, relative to the optical filter (A). The 15 GHz drive applied to the first modulator produces two optical spectral lines at $\Omega \pm \omega$ (B) which are separated by 30 GHz. The optical filter isolates each of these carriers (C and D). One of these carriers is modulated by the second modulator, which is driven by subcarriers of a nominal frequency of 1.5 GHz, leading to the optical spectrum shown in (E). A coupler arrangement combines the unmodulated optical carrier with the modulated optical carrier (F) and samples a small fraction of (E) for detection and use by the frequency control circuitry.

At the antenna unit the incident optical field is detected by the PIN photodiode which leads to a photocurrent spectrum (G) that is amplified before being applied to a diplexer. The unmodulated 30 GHz carrier is recovered (I) and amplified for use as a local oscillator and the upper sideband (H) is selected for transmission. A return signal (generated by a transposer for the demonstrator) at 28 GHz (J) is amplified and downconverted to a 2GHz IF (K). This signal is conveyed back to the base unit by the adoption of WDM technology. At the base unit the signals are recovered by a tuned receiver.

The topology of the proposed 60 GHz broadcast demonstrator is shown in Figure 9.22.

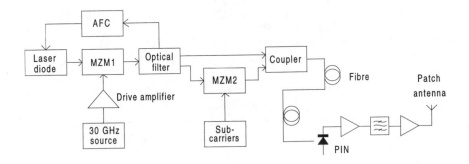

Figure 9.22 *Proposed 60 GHz broadcast demonstrator*

As can be seen from the figure this has the same forward link topology as the 30 GHz bi-directional demonstrator. There are, however, novel realisation problems that have to be addressed before the concept can be extended to 60 GHz operation.

9.4 Concluding remarks

Future mobile communications is expected to make use of mm-wave frequencies, spectral congestion at lower frequencies making these effectively unavailable for broadband applications. The potential of optical fibre techniques to support the generation and remote delivery of mm-wave based mobile broadband services has been outlined, with particular reference to developments effected within the context of the EU RACE programme by project MODAL. It has been shown that such systems can be developed using optical and microwave/mm-wave technology which is readily available today. By employing optical techniques the flexibility for systems deployment is greatly increased, with optical fibre allowing signal transport over tens of kilometers, effecting benefits in base station interconnection and antenna siting and so contributing to low construction, installation and maintenance costs.

9.5 References

1   Goldberg, L; Esman, R.D. and Williams, K.J.: 'Generation and control of microwave signals by optical techniques'. *IEE Proc.-J.*, 1992, **139**(4), pp. 268-294

2   Cooper, A.J.: 'Fibre/radio for the provision of cordless/mobile telephony services in the access network'. *Electron. Lett.*, 1990, **26**(24), pp. 2054-2056

3   Walker, R.G.: 'High speed III-V semiconductor intensity modulators'. *IEEE J. Quantum Electron.*, 1991, **27**(3), pp. 654-667

4   Noguchi, K; Mitomi, O.; Kawano, K. and Yanagibashi, M: 'Highly efficient 40 GHz bandwidth Ti:LiNbO$_3$ optical modulator employing ridge structure'. *IEEE Photonics Technol. Lett.*, 1993, **5**(1), pp. 52-54

5   Sheehy, F.T.; Bridges, W.B. and Schaffner, J.H.: '60 GHz and 94 GHz antenna-coupled LiNbO$_3$ electrooptic modulators'. *IEEE Photonics Technol. Lett.*, 1993, **5**(3), pp. 307-310

6   Harrison, J. and Mooradian, A: 'Linewidth and offset frequency locking of external cavity GaAlAs lasers'. *IEEE J. Quantum Electron.*, 1989, **25**(6), pp. 1152-1155

7   Goldberg, L; Taylor, H.F.; Weller, J.F. and Bloom, D.M.: 'Microwave signal generation with injection locked laser diodes'. *Electron. Lett.*, 1985, **19**(13), pp. 491-493

8   Walker, N.G.; Wake, D. and Smith, I.C.:   'Efficient millimetre-wave signal conversion through FM-IM conversion in dispersive optical fibre links'. *Electron. Lett.*, 1992, **28**(21), pp. 2027-2028

9   Wake, D.; Smith, I.C.; Walker, N.G.; Henning, I.D. and Craver, R.D.:   'Video transmission over a 40 GHz radio-fibre link'. *Electron. Lett.*, 1992, **28**(21), pp. 2024-2025

10 Tun, T.S.; Jungerman, R.L. and Elliot, S.S.: 'Calibration of optical receivers and modulators using an optical heterodyne technique'. *Proc. IEEE MTT-S*, 1988, pp. 1067-1070

11 Kawanishi, S.; Takada, A. and Saruwantatani M.: 'Wideband frequency measurement of optical receivers using optical heterodyne detection'. *IEEE J. of Lightwave Technol.*, 1989, **7**(1), pp. 92-98

12 Williams, K.J.; Goldberg, L.; Esman, R.D.; Dagerais, M. and Weller, J.F.: '6-34 GHz offset phase locking of Nd:YAG 1319 nm nonplanar ring lasers'. *Electron. Lett.*, 1989, **25**(18), pp. 1242-1243

13 Gliese, U.; Nielsen, T.N.; Bruun, M.; Christiensen, E.L.; Stubkjaer, K.E.; Lindgren, S. AND Broberg, B.:   'A wideband heterodyne optical phase-locked loop for generation of 3-18 GHz microwave carriers'. *IEEE Photonics Technol. Lett.*, 1992, **4**(8), pp. 936-938

14 Gallion, P.B. and Debarge, G.: 'Quantum phase noise and field correlation in single frequency semiconductor laser systems'. *IEEE J. Quantum Electron.*, 1984, **20**(4), pp. 343-349

15 Schmuck, H.; Pfeiffer, Th. and Bülow, H.: 'Design optimisation of erbium ring laser regarding output power and spectral properties'. *Electron. Lett.*, 1992, **28**(17), pp. 1637-1639

Chapter 10

Optical receiver design for optical fibre SCM systems

P. M. R. S. Moreira, I. Z. Darwazeh and J. J. O'Reilly

10.1 Introduction

The use of optical fibres to provide multichannel distribution and broadcast services, to a large number of users, is now a field of intense interest [1,2]. Several strategies have been proposed for this purpose [3,4,5] and one of the more attractive strategies for analogue and mixed analogue/digital information services is the one whereby each information channel is used to amplitude-, frequency- or phase-modulate a separate RF carrier. Linear addition of the modulated carriers results in a composite frequency division multiplexed (FDM) signal that is used to intensity-modulate a laser diode. The RF carriers are referred to as "subcarriers", so they can be distinguished from the main optical carrier, and the resulting multiplexing strategy is termed subcarrier multiplexing (SCM) [5,6]. Alternatively, each of the modulated subcarriers can be used to intensity-modulate a separate laser diode with the outputs of all laser diodes being summed optically before transmission down a fibre [5]. In both cases, at the receiver end, following optoelectronic conversion and recovery of the FDM signal, standard downconversion and tuning techniques are used to recover the baseband information of a user-selected channel.

SCM systems have been extensively studied [5,6,7] with successful laboratory demonstrations [8,9] and field trials for video distribution [10] reported. They have several attractive features when compared to other multiplexing schemes. One of these is that they allow easy selection of a certain information channel by the end user. Also, the user terminal receiver equipment is somewhat simpler than that of a time division multiplexed (TDM) system, since high speed multiplexing, demultiplexing and synchronisation circuitry are not needed and a narrowband (as opposed to wideband) receiver front end can be employed, resulting in lower receiver noise and consequently better signal quality.

10.1.1 SCM system performance and relation to receiver noise

At the end terminal, SCM signals are objectively assessed in terms of carrier to noise ratio (CNR) of the detected and amplified FDM signal or one of its component channels. Taking the case of an N channel system and defining an optical modulation depth m as:

$$m = \frac{P_{rf}}{P_{dc}} \tag{10.1}$$

where P_{rf} and P_{dc} are the half peak to peak optical signal power and the average optical power of a single optical source, respectively. The average photocurrent generated by a pin photodiode, having a responsivity $\Re$, is given by:

$$I_a = N P_{dc} \Re \tag{10.2}$$

and the rms signal photocurrent is given by:

$$I_p = \frac{1}{\sqrt{2}} m I_a \tag{10.3}$$

At the output of the optical receiver preamplifier, the CNR of a single channel can be given by the simplified expression:

$$CNR = \frac{m^2 I_a^2}{2 B \left(S_{rx} + S_{sh} + S_{RIN} \right)} \tag{10.4}$$

where B is the channel bandwidth, S_{rx}, S_{sh} and S_{RIN} are the noise spectral density contributions of the receiver circuit noise, the optical signal dependent shot noise and the optical source relative intensity noise (RIN), respectively. All contributions are taken with reference to the input of the receiver electronic pre-amplifier.

Darcie [5] concludes that circuit and shot noise contributions are dominant in systems employing FM or digital signalling schemes, while RIN noise contribution has to be considered in systems employing AM-VSB modulation. Other noise contributions relating to system noise, channel crosstalk and intermodulation distortion have also to be considered for accurate assessment of SCM systems performance[1].

Equation 10.4 above indicates that the CNR quality can be improved by reducing the receiver circuit noise contribution. SCM receivers are either circuit noise limited or signal (mainly shot and RIN) noise limited. In the cases where RIN noise does not dominate the noise performance it is important to minimise the electronic receiver

---

[1] Refer, for instance, to chapter 7 in this book.

noise so that the shot noise limit can be approached. Noise minimisation, achieved through receiver preamplifier circuit tuning, is the subject of discussion in the following parts of this chapter.

10.2 Noise in optical receivers

Before the question of noise minimisation is considered, noise sources in optical receivers are discussed below, followed by an outline of an appropriate method of noise representation.

In a pin-FET optical receiver two noise mechanisms contribute to the degradation of SNR performance. The first of these is shot noise generated by the photocurrent. This source has a noise power spectral density (NPSD) given by[2] :

$$\frac{d}{df}\langle i_s\, i_s^*\rangle = q\, I_a \qquad \left(A^2/Hz\right) \qquad\qquad (10.5)$$

where q is the electronic charge and * denotes the complex conjugate.

This noise term is a function of the system configuration and components, and cannot be modified by the receiver designer. The second noise generating mechanism is thermal noise contributed by passive and active receiver elements. These noise sources can, to a limited extent, be controlled by the circuit designer if appropriate circuit topologies, circuit elements and noise tuning techniques are employed.

In the receiver thermal noise is generated by all lossy passive elements. The associated NPSD given by:

$$\frac{d}{df}\langle i_{th}\, i_{th}^*\rangle = \frac{2k\theta}{R} \qquad \left(A^2/Hz\right) \qquad\qquad (10.6)$$

where k is Boltzman's constant, θ is the absolute temperature in Kelvin and R is the element loss resistance.

SCM receivers are normally designed to operate in the multi-GHz regime where MESFET or HEMT based circuits can achieve the lowest noise figures. For such devices noise modelling is carried out by separating the noisy FET into two parts, the extrinsic and the intrinsic device. The extrinsic noisy FET, shown in Figure 10.1, consists of the noiseless elements L_g, L_s, L_d and C_{ds} and the access resistances R_g, R_s, and R_d that act as thermal noise generators with NPSD given by Equation 10.6. The reactive elements L_g, L_s, L_d and C_{ds}, whilst not contributing directly to the generation of noise, introduce a frequency dependence in the noise generated by the

---

[2] Double sided power spectral densities are used throughout this chapter.

extrinsic FET thermal noise sources and transform the noise spectrum associated with the intrinsic device.

The model chosen to represent the intrinsic noisy FET is shown in Figure 10.2. This model has been extensively used in the literature and was first proposed by Van der Ziel [11, 12]. The model characterises the noise behaviour of the intrinsic FET by introducing two noise generators: one in the gate input circuit i_g and the other in the drain output circuit i_d. The resistors R_i and R_{ds} are assumed to be noiseless, with their noise contribution totally included in i_g and i_d.

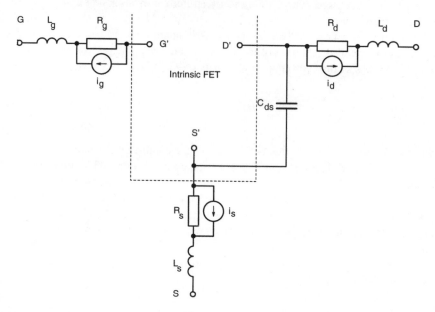

Figure 10.1 *Extrinsic FET noise model*

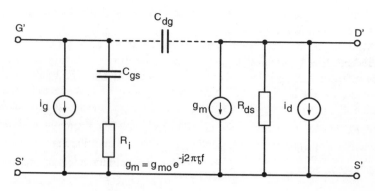

Figure 10.2 *Intrinsic FET noise model*

The noise current generator i_d represents the short circuit channel noise generated in the drain-source path. This can be shown to have a white spectrum in the microwave frequency range [13] with NPSD given by

$$\frac{d}{df}\langle i_d\, i_d^* \rangle = 2\,k\theta\, g_{m0}\, P \tag{10.7}$$

where P is a dimensionless coefficient that depends on the d.c. bias conditions, the device geometry and the technological parameters.

The total induced gate charge fluctuation is described, in the intrinsic noise model, by the noise current generator i_g which has a frequency dependent NPSD given by

$$\frac{d}{df}\langle i_d\, i_d^* \rangle = 2\,k\theta\, \frac{\left(\omega\, C_{gs}\right)^2}{g_{m0}}\, R \tag{10.8}$$

where $\omega = 2\pi f$ and R is a dimensionless coefficient also dependent on bias and process parameters. Since i_g and i_d have a common physical origin, they are partially correlated with cross-correlation spectral density given by

$$\frac{d}{df}\langle i_g\, i_d^* \rangle = j\,2\,k\theta\, \omega\, C_{gs}\, C\,\sqrt{P\,R} \tag{10.9}$$

where C is the correlation coefficient. As is the case for the other two coefficients C is a function of technological parameters, the bias conditions and device geometry. It is of interest to note that the cross-correlation spectral density is purely imaginary since it results from capacitive coupling of the gate circuit with the drain circuit noise sources.

This model can be further extended by considering other noise mechanisms such as flicker noise [13,14] and gate leakage current noise [13,15]. However, these noise sources can be ignored for SCM receivers, since they are typically operated at frequencies beyond the region where such noise mechanisms are significant.

In order to evaluate the receiver CNR performance, given by Equation 10.4, all the electronic noise contributions must be referred to the electronic preamplifier input. Several noise analysis methods can be used for this purpose [16-19]. In this work the method followed was that proposed by Hilbrand and Russer [19], based on the description of noise by means of correlation matrices. This method is well suited to the type of networks encountered in optical receivers operating at microwave frequencies and can be simply and efficiently used for computer-aided noise analysis. A brief description of this technique is given in the Appendix at the end of this chapter.

10.3 Noise minimisation

In this section the problem of noise minimisation in optical receivers will be addressed. The most commonly adopted techniques for reducing circuit noise will be discussed by considering the case of an optical receiver that uses a generic type of noise-matching network between the photodiode and the front-end amplifier. The treatment is general and two specific results, valid for broadband and narrowband receivers, are derived. First, an expression is given that relates the optical receiver equivalent input noise current spectral density with the familiar microwave description of circuit noise in terms of the parameters F_{min}, Y_{opt} and R_n. Second, the criteria for optimal noise matching in optical receivers are established.

10.3.1 Noise-matched receivers

Optimisation of digital optical receivers is usually couched in terms of choosing the optimum impulse response leading to the minimum bit error rate (BER) for a given optical power [20, 21]. On the other hand, analogue receivers are optimised in terms of a transfer function chosen to attain flat gain and linear phase responses across the band of interest, with minimum added noise and distortion. For both types of receivers electronic noise is considered as a Gaussian random variable with variance given by

$$\sigma_c^2 = \int_{-\infty}^{+\infty} S_{eq}(f) |H_T(f)|^2 df \tag{10.10}$$

where $S_{eq}(f)$ is the equivalent noise power spectral density (NPSD) at the receiver input and $H_T(f)$ is the receiver transfer function.

The variance, σ_c^2, should ideally be minimised in order to maximise the CNR of an analogue receiver or minimise the sensitivity figure of a digital receiver. This variance can be minimised by the appropriate choice of $H_T(f)$. However, not only noise but also amplitude and phase distortions in analogue applications, or intersymbol interference, tolerance to jitter and so ultimately sensitivity in digital applications, all depend on the receiver $H_T(f)$ and it is thus not possible to choose $H_T(f)$ based solely on noise considerations. However, the value of σ_c^2 can also be minimised by using appropriate circuit techniques to reduce the value of the power spectral density $S_{eq}(f)$ of the equivalent input noise current. Some of the well-known design solutions that can be used to reduce the noise generated in the optical receiver include the following:

- selection of the best circuit configuration;
- careful selection of the active devices;
- selection of the optimum bias conditions for the active devices;
- use of a noise-matching network.

All these solutions can be discussed in a unified manner by considering the general case of an optical receiver that uses a noise-matching network between the photodiode and the front-end. The concept of noise-matching for optical receivers was first proposed by Hullet and Muoi [22] who showed that an inductance in series with the front-end input can improve the receiver's noise performance. The concept has been adopted for very high bit rate optical receivers [23] where more complex noise matching networks have been used to produce the lowest equivalent input noise current spectral densities reported [24-27].

The generic noise model of an optical receiver employing a noise-matching network is represented in Figure 10.3.

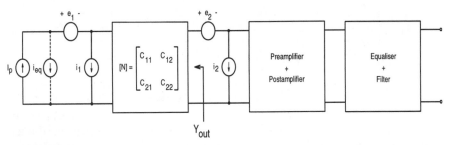

Figure 10.3 *Optical receiver noise model*

Here, I_p is the signal current and N is a two port network that, without loss of generality, represents the photodiode equivalent circuit (excluding the signal current source), the noise matching network and the photodiode bias network, e_1 and i_1 are the noise sources that characterise the noise behaviour of the network N while e_2 and i_2 characterise the noise behaviour of the active part of the optical receiver. The noise sources e_2, i_2, e_1 and i_1 can be replaced by two equivalent noise sources e and i at the input of the network N. In the analysis, all the noise sources will be considered to be characterised by their correlation matrices, as discussed in the appendix, and the chain-matrix will be the adopted representation unless otherwise stated. Noise sources e_2, i_2 can be referred to the input and their effect lumped with that of noise sources e_1 and i_1 using the transformation of Equation 10.58

$$\mathbf{C}_c = \mathbf{C}_{c1} + \mathbf{N}\,\mathbf{C}_{c2}\mathbf{N}^+ \tag{10.11}$$

with

$$\mathbf{C}_c = \begin{bmatrix} \dfrac{d}{df}\langle e\,e^*\rangle & \dfrac{d}{df}\langle e\,i^*\rangle \\[2ex] \dfrac{d}{df}\langle e^*\,i\rangle & \dfrac{d}{df}\langle i\,i^*\rangle \end{bmatrix} \tag{10.12}$$

$$\mathbf{N} = \begin{bmatrix} c_{11} & c_{12} \\ c_{21} & c_{22} \end{bmatrix} \tag{10.13}$$

$$\mathbf{C}_{c1} = \begin{bmatrix} \dfrac{d}{df}\langle e_1\,e_1^*\rangle & \dfrac{d}{df}\langle e_1\,i_1^*\rangle \\[2ex] \dfrac{d}{df}\langle e_1^*\,i_1\rangle & \dfrac{d}{df}\langle i_1\,i_1^*\rangle \end{bmatrix} \tag{10.14}$$

$$\mathbf{C}_{c2} = \begin{bmatrix} \dfrac{d}{df}\langle e_2\,e_2^*\rangle & \dfrac{d}{df}\langle e_2\,i_2^*\rangle \\[2ex] \dfrac{d}{df}\langle e_2^*\,i_2\rangle & \dfrac{d}{df}\langle i_2\,i_2^*\rangle \end{bmatrix} \tag{10.15}$$

where $\mathbf{C}_c$ is the correlation matrix of the total network, $\mathbf{C}_{c1}$ and $\mathbf{N}$ are respectively the correlation and chain-matrix of the noise matching-network, and $\mathbf{C}_{c2}$ is the correlation matrix of the amplifier. Both noise source e and i can be further combined into the equivalent input noise current i_{eq}. However, since the signal current source I_p effectively represents an open circuit to the voltage noise source e, the equivalent input noise current i_{eq} is simply equal to i. Then, from Equation 10.11 the noise spectral density of the equivalent input noise current i_{eq} is given by:

$$S_{eq}(f) = \frac{d}{df}\langle i_{eq}\,i_{eq}^*\rangle = \frac{d}{df}\langle i\,i^*\rangle \tag{10.16}$$

with

$$\frac{d}{df}\langle i\,i^*\rangle = \frac{d}{df}\langle i_1\,i_1^*\rangle + |c_{21}|^2\,\frac{d}{df}\langle e_2\,e_2^*\rangle + 2\,\mathrm{Re}\!\left(c_{21}\,c_{22}^*\,\frac{d}{df}\langle e_2\,i_2^*\rangle \right) + |c_{22}|^2\,\frac{d}{df}\langle i_2\,i_2^*\rangle \tag{10.17}$$

In this equation the first term can be recognised as the noise power spectral density generated by the noise-matching network (including the photodiode and input bias circuit) while the remaining terms correspond to the noise power and cross-correlation spectral densities due to the amplifier, transformed by the presence of the network N.

10.3.2 Front end selection

From the discussion above, it is clear that only one element of the total correlation matrix $\mathbf{C}_c$ in Equation 10.11 is used to estimate the total input noise current. Although the other elements of this matrix are not directly relevant to the calculation of $S_{eq}(f)$ they can provide insight into the question of selection of the best front-end configuration from the point of view of noise. That is, in Equation 10.11 the input network can be thought of as the first amplifier stage with chain-matrix $\mathbf{N}$ and noise correlation-matrix $\mathbf{C}_{c1}$, while $\mathbf{C}_{c2}$ is the noise-correlation matrix of the following amplifier stages. The merits of a given front-end topology must then be evaluated under two criteria. First, the contribution of the stage itself to the receiver noise, i.e. the magnitude of the elements of $\mathbf{C}_{c1}$, must be considered. Second, the capacity of the stage to reduce the contribution of the following stages to the total receiver noise must be taken into account. It is a widely known result that for the three possible configurations – common-source (CS), common-drain (CD) and common-gate (CG) – the magnitude of the elements of the noise-correlation matrix $(\mathbf{C}_{c1})$ are almost equal [28-30]. Accordingly, the choice of one of the three circuit configurations has to be based on the second criterion, the ability of the circuit to minimise the noise contribution of the subsequent amplifier stages as well as the contribution of its own load. As given by Equation 10.11 the matrix $\mathbf{C}_{c2}$,when referred to the input of network N, is transformed by its chain parameters and each element of the transformed matrix is a weighted sum, by the elements of $\mathbf{N}$, of the original elements of $\mathbf{C}_{c2}$. Since the CS configuration has the largest transfer parameters – or, equivalently, the smallest chain parameters – it should thus be selected as the front-end stage whenever the best noise performance is to be achieved. On the other hand the CG configuration has the smaller transfer parameters of the three configurations and is thus the least efficient in reducing the noise contribution of the subsequent stages [28, 29].

10.3.3 Optimum noise matching

Returning now to Equation 10.17 and assuming that the noise contribution of the subsequent amplifying devices is minimised by proper selection and design of the first stage, it is usually a good approximation to consider the photodiode, the noise- matching network and the front-end device as the dominant noise sources in the circuit. In this case, the noise terms involving noise generators e_2 and i_2 in Figure 10.3 can be considered to be due to the first active device only.

As discussed in [19], the power and cross-correlation spectral densities of these generators can be related to the device noise parameters by:

$$\mathbf{C}_{c2} = 2k\theta_0 \begin{bmatrix} R_n & \dfrac{F_{min}-1}{2} - R_n Y_{opt} \\ \dfrac{F_{min}-1}{2} - R_n Y_{opt}^* & R_n |Y_{opt}|^2 \end{bmatrix}$$

(10.18)

where θ_0 is the standard reference temperature ($\theta_0 = 290\,\text{K}$), R_n is the noise resistance, Y_{opt} is the optimum source admittance and F_{min} is the minimum noise factor. If the power and cross-correlation spectral densities, as expressed by Equation 10.18, are replaced in Equation 10.17, this equation can now be rewritten as:

$$S_{eq}(f) = \frac{d}{df}\langle i_1 i_1^* \rangle + 2k\theta_0 |c_{22}|^2 \left[R_n |Y_{out} - Y_{opt}|^2 + (F_{min}-1)\text{Re}(Y_{out}) \right]$$

(10.19)

where $Y_{out} = c_{21}/c_{22}$ is the output admittance of the network N. Since the noise-matching network is passive the first term in Equation 10.19 can be computed from its $\mathbf{Z}$ or $\mathbf{Y}$ matrix. Assuming that the two-port $\mathbf{Z}$-matrix of the network N is defined – as is generally the case due, at least, to the presence of the shunt photodiode capacitance – then its correlation matrix in the chain representation is given by Equations 10.61 and 10.59 in the Appendix.

$$\mathbf{C}_{c1} = 2k\theta\, \mathbf{T}\,\text{Re}(\mathbf{Z})\mathbf{T}^+$$

(10.20)

where θ is the network temperature and $\mathbf{T}$ is the transformation matrix between the impedance and the chain representations given in Table 10.2 in the Appendix. From this:

$$\frac{d}{df}\langle i_1 i_1^* \rangle = 2k\theta |c_{21}|^2 \,\text{Re}(z_{22})$$

(10.21)

Since $z_{22} = Z_{out} = 1/Y_{out}$ and $Y_{out} = c_{21}/c_{22}$ Equation 10.21 can be rewritten as:

$$\frac{d}{df}\langle i_1 i_1^* \rangle = 2k\theta |c_{22}|^2 \,\text{Re}(Y_{out})$$

(10.22)

Finally, by substituting this in Equation 10.19, the optical receiver equivalent input noise current power spectral density is:

$$S_{eq}(f) = 2k\theta_0 |c_{22}|^2 Y_N(f)$$

(10.23)

where $Y_N(f)$ will be defined here as the equivalent noise admittance of the total network and is given by:

$$Y_N(f) = \frac{\theta}{\theta_0} \text{Re}(Y_{out}) + R_n \left| Y_{out} - Y_{opt} \right|^2 + (F_{min} - 1) \text{Re}(Y_{out}) \qquad (10.24)$$

From Equations 10.23 and 10.24 the criteria for optimum noise-matching can now be derived. These equations indicate that the equivalent input noise current spectral density can be reduced by maximising the current gain factor $1/c_{22}$ and by minimising the optical receiver equivalent noise admittance $Y_N(f)$. The noise admittance (Equation 10.24) is a function of the output admittance of the noise-matching network N and as such its value can be minimised by proper design of Y_{out}.

If Y_{out} is written as $Y_{out} = G_o + j X_o$ then its optimum value can be obtained by solving the following two partial differential equations:

$$\frac{\partial}{\partial G_o} Y_N(f) = 0 \qquad (10.25)$$

$$\frac{\partial}{\partial X_o} Y_N(f) = 0 \qquad (10.26)$$

for G_o and X_o. From these equations the optimum admittance is

$$G_o = \text{Re}(Y_{opt}) + \frac{1}{2 R_n}\left[1 - \frac{\theta}{\theta_0} - F_{min} \right] \qquad (10.27)$$

$$X_o = \text{Im}(Y_{opt}) \qquad (10.28)$$

These show that – contrary to what as been suggested in [25, 27] – the optimum noise-matching admittance for optical receivers differs from the optimum noise figure matching that requires $Y_{out} = Y_{opt}$. It should be noticed, however, that for some devices, Equation 10.27 can result in a negative real part of the optimum admittance contradicting the realisability condition for a passive network. Nonetheless, Equations 10.27 and 10.28 correspond to a single minimum of $Y_N(f)$ and whenever G_o, as given by Equation 10.27, becomes negative the optimum noise matching condition for optical receivers can be restated as:

$$G_o = 0 \qquad (10.29)$$

$$X_o = \text{Im}(Y_{opt}) \qquad (10.30)$$

with corresponding minimum equivalent noise admittance:

$$Y_N(f) = R_n \, \text{Re}(Y_{opt})^2 \tag{10.31}$$

It is interesting to notice that this corresponds to the widely used high impedance approach [31-33] to receiver noise minimisation plus an additional condition that states that the front-end should see an admittance $Y_{out} = j \, \text{Im}(Y_{opt})$. Figure 10.4 represents the equivalent noise admittance $Y_N(f)$ for a $4 \times 75 \, \mu m$ commercially available GaAs MESFET in three idealised noise-matching situations: with no matching, with matching to Y_{opt} and with matching to $j \, \text{Im}(Y_{opt})$.

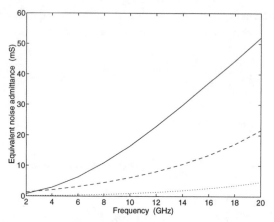

Figure 10.4 *Equivalent noise admittance for three matching situations: no matching (solid line), matching to Y_{opt} (broken line) and matching to $j \, \text{Im}(Y_{opt})$ (dotted line)*

It can be observed from this figure that, although matching to Y_{opt} already provides an improvement when compared with the unmatched case, matching to $j \, \text{Im}(Y_{opt})$ further reduces the value of the equivalent noise admittance $Y_N(f)$. As mentioned before, minimisation of $S_{eq}(f)$ amounts to the simultaneous maximisation of the current gain factor $1/c_{22}$ and minimisation of $Y_N(f)$. However, these are interrelated and consequently Y_{out} and c_{22} cannot be chosen independently. The choice of the optimum set of chain parameters for the matching network is thus best made numerically by solving the constrained optimisation problem

$$\min \left\{ \sigma_c^2 = \int_{-\infty}^{+\infty} S_{eq}(f) |H_T(f)|^2 \, df \right\} \tag{10.32}$$

subject to the constraints

$$\left| Y_{out} - j\,\text{Im}\!\left(Y_{opt}\right) \right| < \varepsilon \qquad \text{with} \qquad Y_{out} = \frac{c_{21}}{c_{22}} \tag{10.33}$$

$$\text{Re}(c_{12})\text{Re}(c_{22}) + \text{Im}(c_{12})\text{Im}(c_{22}) \geq 0 \tag{10.34}$$

$$\text{Re}(c_{11})\text{Re}(c_{12}) + \text{Im}(c_{11})\text{Im}(c_{12}) \geq 0 \tag{10.35}$$

$$\text{Re}(c_{12})\text{Re}(c_{21}) + \text{Im}(c_{11})\text{Im}(c_{22}) \geq 0 \tag{10.36}$$

where (10.33) allows a certain deviation from the optimum matching condition to accommodate for losses in the matching network and the photodiode parasitics, while constraints (10.34) to (10.36) ensure that the matching network is realisable[3] [34]. Once the optimum set of the chain parameters is determined, a network can be designed that approximates the desired response. Alternatively, a circuit topology can be chosen at the outset and its elements optimised to provide close to optimum noise-matching over the frequency range of interest. This is the most common approach used in the literature – with the exception of the work of Park and Minasian [25, 27] where a matching network was designed to realise the minimum noise figure matching condition $Y_{out} = Y_{opt}$ – and it will be used in the next section for the noise optimisation of an analogue receiver for subcarrier multiplexed applications.

10.3.4 Active device selection and optimum bias conditions

The problem of selecting an optimum active device and its optimum bias condition is probably the most complex one and can be approached from the device design view point [35-37]. However, the circuit designer is normally faced with a limited choice of technologies and devices. In this case, Equation 10.24 can guide the designer's selection. For instance, in situations where optimum noise. matching can be realised, Equation 10.24 indicates that devices and bias conditions leading to the lowest values of $R_n\,\text{Re}(Y_{opt})^2$ should be selected while, for example, in a high impedance approach where the optimum noise matching condition is not satisfied a low value of R_n is preferred

---

[3] Here it is assumed that the noise-matching is not significantly affecting the gain of the first stage at any frequency within the range of interest. If this is not so, then either condition (10.33) has to be relaxed and a multi-objective optimisation strategy adopted that will ensure that the first stage gain does not drop below a certain level or, alternatively, the noise of subsequent stages must be considered in Equation 10.24. Under such circumstances the noise parameters in this equation represent the first and subsequent stages and not only the first stage as is being assumed here.

10.4 SCM GaAs receiver with integrated tuning network

Most SCM systems proposed to date are to be operated with subcarrier frequencies in the microwave region. For such systems to be widely used, reliable, economic and easy to produce, system components have to be made available. For operation in the GHz region, receiver implementation as a microwave monolithic integrated circuit (MMIC) will satisfy the requirements of reliability and reproducibility and will be economically competitive when compared to its discrete or hybrid equivalent.

The concepts discussed in the previous section will now be applied to the design and optimisation of a noise-tuned SCM optical receiver integrated as a GaAs MMIC that includes an integrated noise-matching network [38-40]. A critical study of tuning networks suitable for integration is undertaken and details of the design and optimisation strategy for such networks are discussed. Simulation results predict low noise and high gain performance, demonstrating the appropriateness of the design technique adopted.

10.4.1 Noise tuning networks

Several types of tuning networks have been examined in the literature for broadband [23], coherent [41] and SCM [42] optical receivers. Since SCM receivers are usually restricted to narrow-band operation [42] a wide choice of low-pass and band-pass networks can be used to realise noise-matching of the front-end over the frequency range of operation [41, 43]. The most common topologies used for such applications are inductive parallel and serial tuning [42], T and Π tuning [41]. The microstrip T-equivalent of transformer tuning has also been used [44], however this type is not suitable for integration in the low GHz frequency range where the quarter wavelength on GaAs substrates exceeds 10 mm. Consequently, only the first four types of tuning network will be considered here for integration with the receiver.

Before the tuning networks can be analysed in detail it is necessary to consider how the tuning elements' non-idealities affect the four basic topologies. For MMIC structures, such as those studied here, this can be done by considering the equivalent circuit models of the GaAs inductors and capacitors which are the basic building blocks of the tuning networks. Measurement-based equivalent circuit models for integrated passive components are usually provided by MMIC manufacturers [45]. Although such models are accurate up to high frequencies, they are too complex to be used in the initial analysis and will be replaced here by the simplified models shown in Figure 10.5; full models will be used in the final design stages.

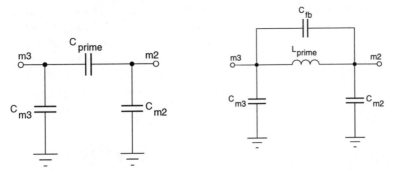

Figure 10.5 *Simplified GaAs MMIC capacitor (left) and inductor (right) models*

Figure 10.6 shows the simplified GaAs equivalent circuits for the shunt, series, T and Π tuning networks when the ideal capacitors and inductors in these networks are replaced by the GaAs models in Figure 10.5. In Figures 10.6(a) and (d) the shunt parasitic capacitances of the GaAs inductors at the input and output ports were excluded from the tuning network and in the analysis they will be considered as being part of the photodiode and amplifier input capacitance respectively. In this way, each branch in the tuning network can be considered as a parallel combination of an ideal inductor and capacitor which in the analysis will be considered as the tunable elements. In a practical implementation each of these elements can be replaced by a combination of GaAs components, including its parasitics.

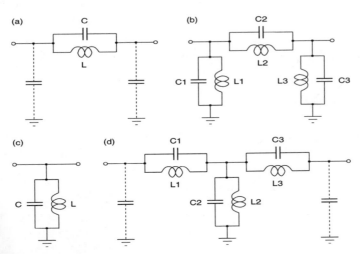

Figure 10.6 *Simplified models for the GaAs implementation of: (a) series tuning, (b) Π tuning, (c) shunt tuning and (d) T tuning*

The equivalent input noise spectral density $S_{eq}(f)$ can now be calculated for an optical receiver which uses any one of the four tuning networks in Figure 10.6 using Equation 10.17. To derive the specific equations for each of the tuning networks it will be further assumed that no thermal noise is generated in the photodiode and that its equivalent circuit consists only of the junction capacitance. In this case, since the tuning networks are considered to be lossless, the first term in Equation 10.17 is zero. The other terms in this equation are, however, non-zero and $S_{eq}(f)$ for the four networks of Figure 10.6 can be expressed as:

$$S_{eq}(f) = |A|^2 \left[|B|^2 \frac{d}{df} \langle e_2 \, e_2^* \rangle + 2 \operatorname{Re}\left(B^* C \frac{d}{df} \langle e_2 \, i_2^* \rangle \right) + |C|^2 \frac{d}{df} \langle i_2 \, i_2^* \rangle \right] \quad (10.37)$$

For the series and Π-network

$$A = \frac{1}{Z_1 Z_3} \quad (10.38)$$

$$B = Z_1 + Z_2 + Z_3 \quad (10.39)$$

$$C = (Z_1 + Z_2) Z_3 \quad (10.40)$$

and for the shunt and T-network:

$$A = \frac{1}{Z_d Z_2} \quad (10.41)$$

$$B = \frac{(Z_d + Z_1)(Z_2 + Z_3) + Z_2 Z_3 + Z_{in}(Z_d + Z_1 + Z_2)}{Z_{in}} \quad (10.42)$$

$$C = (Z_d + Z_1)(Z_2 + Z_3) + Z_2 Z_3 \quad (10.43)$$

In the above equations,

$$\frac{1}{Z_i} = j\omega C_i + \frac{1}{j\omega L_i} \qquad i = 1, 2, 3 \quad (10.44)$$

$$Z_d = \frac{1}{j\omega C_d} \quad (10.45)$$

$$Z_{in} = \frac{1}{j\omega C_{in}} \quad (10.46)$$

where C_d is the photodiode capacitance and C_{in} is the preamplifier input capacitance. For the case of the Π network the capacitances C_1 and C_3 include the photodiode and the preamplifier input capacitance, respectively, and in the case of series tuning $C_1 = C_d$, $C_3 = C_{in}$.

Equation 10.37 was derived assuming a lossless tuning network and ignoring the thermal noise generated by the bulk resistance of the photodiode. Such assumptions serve to simplify the initial design process and provide a first measure of the effectiveness of each of the tuning networks to minimise the noise generated by the amplifier alone. They can also be used to obtain a first estimate of the values of the GaAs L's and C's to be used in the receiver design by appropriately choosing those element values that minimise the equivalent input noise spectral density $S_{eq}(f)$ as given by Equation 10.37. However, to complete the design, detailed noise models that account for all the noise sources in the amplifier, losses in the tuning network and the noise generated in the photodiode itself need to be used. This process of going from Equation 10.37 to a full operational noise-optimised optical receiver will now be described taking into consideration the specific requirements of SCM systems and the practical limitations imposed by implementation on GaAs. Also, a critical comparison of four possible receiver topologies is undertaken through study of a set of simulation results.

10.4.2 Optimisation procedure

Tuned receivers used for SCM systems typically fall into one of two classes: narrow- and wide-band. In the first class a mixer is used prior to amplification to select a single channel while in the wide-band class all channels are simultaneously amplified then post-amplification mixing is carried out for channel selection. The second of these two classes imposes more stringent design requirements – it is this type which is studied here.

The choice of the optimum component values for a given tuning network depends primarily on the network's ability to maximise the CNR for all the received channels. It will be assumed here that all the channels are narrowband and that the receiver operation is circuit noise limited. Under these conditions optimisation of a given channel CNR amounts to minimisation of the receiver equivalent input noise spectral density at the channel carrier frequency. Since several of these channels are to be detected and amplified by the receiver, it is important to keep the value of $S_{eq}(f)$ as low as possible over the receiver bandwidth. To achieve this objective the following cost function was selected to control the optimisation process:

$$E = \lambda \cdot D + I \tag{10.47}$$

where, for the tuning range f_{min} to f_{max},

$$I = \frac{1}{f_{max} - f_{min}} \int_{f_{min}}^{f_{max}} S_{eq}(f)\,df \tag{10.48}$$

$$D = \int_{f_{min}}^{f_{max}} \left| S_{eq}(f) - I \right| df \tag{10.49}$$

and λ is a weighting factor. This cost function constrains the optimiser to find a solution that corresponds to a low mean value of $S_{eq}(f)$ (Equation 10.48) and avoids solutions where some channels might suffer from poor SNR (Equation 10.49) as is the case of trace (b) in Figure 10.7.

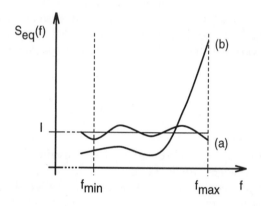

Figure 10.7 *Two possible outcomes of $S_{eq}(f)$ optimisation: (a) conditions set by Equations 10.48 and 10.49 considered and (b) condition set by Equation 10.49 ignored*

Initial optimisation of the tuning network can be carried out for $S_{eq}(f)$, as defined in Equation 10.37, with constraints applied to component values limiting them to those allowed by the MMIC process used. At this stage, only the noise contribution of the first MESFET is included. From this initial optimisation stage component values are obtained and then used for the generation of the initial MMIC layout. Further optimisation is then carried out using a circuit analysis program with full equivalent models used for all circuit components, accounting for the noise contributions of all the amplifier stages and the thermal noise generated in the tuning network and photodiode. During the optimisation process, the active device noise model described in this chapter using de-embedded values of P, R and C [17, 46] can be used in the numerical simulation of the receiver noise and to calculate the power and cross-correlation spectral densities in Equation 10.37.

10.4.3 Design example

In this section the design of a narrowband SCM tuned receiver, integrated with its noise-tuning network as a GaAs MMIC, is discussed. The design is optimised to operate over the frequency range $1.8 - 2.2\,\text{GHz}$. Different tuning networks are examined and their performance optimised, following the procedure outlined in Section 10.4.2, under the constraints imposed by practical implementation aspects of the GaAs process considered. Results of the noise and frequency response for the receiver are presented.

The optimised input noise equivalent spectral densities for a three-stage GaAs receiver preamplifier (described in the next section), when preceded by each of the four types of tuning network, are presented in Figure 10.8.

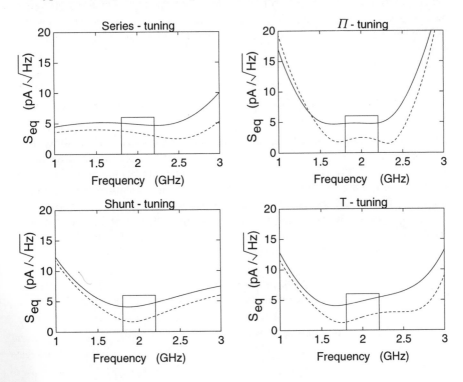

Figure 10.8 *Optimised $S_{eq}(f)$ for different tuning networks: full GaAs models (solid line) and Equation 10.37 (broken line)*

The corresponding optimised component values are given in Table 10.1.

Table 10.1 *Optimised component values*

Network	Component					
	L_1 (nH)	C_1 (pF)	L_2 (nH)	C_2 (pF)	L_3 (nH)	C_3 (pF)
T	5.42	0.08	6.08	0.06	2.75	0.61
Π	12.07	0.11	11.85	0.06	11.48	0.06
Series	12.6	-	-	-	-	-
Shunt	7.34	-	-	-	-	-

For all cases the tuning range was set from 1.6 to 2.4 GHz. This is wider than the proposed operating range (1.8 to 2.2 GHz). This choice was made to provide a more robust receiver, avoiding the possible exposure of channels located near the band edges to higher noise levels, in case the circuit parameters or the circuit response varied slightly due to process or circuit variations. When no tuning network is used the input noise spectral density varied, over the frequency band of interest, from 7.5 to $9.5\,\mathrm{pA}/\sqrt{\mathrm{Hz}}$. However, the use of a tuning circuit reduced this noise to less than $6\,\mathrm{pA}/\sqrt{\mathrm{Hz}}$ over the whole optimisation range and to less than $5\,\mathrm{pA}/\sqrt{\mathrm{Hz}}$ over the receiver operating range, as shown in the graphs. The figures also contrast the results of the two phases of optimisation. From these, it is clear that when ideal components are used, the best results are obtained for the T and Π networks. However, when full GaAs equivalent models are used, these two networks were more severely affected by the process non-idealities than were the simpler series and shunt networks. Also, they show faster rise in noise outside the optimisation range. When comparing the behaviour of the series and shunt tuning networks it can be seen that there is a slight noise advantage for the shunt tuning case. However, the series tuning results display a flatter noise spectral density over the optimisation range.

Figure 10.9 compares the transimpedance gains of the receiver for the different tuning networks. For all cases a midband transimpedance higher than $70\,\mathrm{dB}\Omega$ is predicted. The series, Π and T networks show reasonably flat gain from 1.8 to 2.2 GHz, while the shunt tuning shows a gain variation of $8\,\mathrm{dB}$.

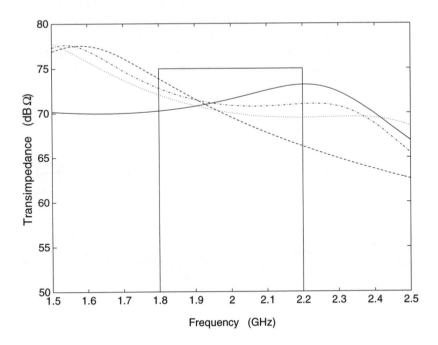

Figure 10.9 *Transimpedance gain for different tuning networks: series tuning (solid line), shunt tuning (broken line), Π- tuning (dashdot line) and T- tuning (dotted line)*

10.4.4 Receiver preamplifier circuit details

The well-behaved input noise and transimpedance characteristics of the series tuning, together with its simplicity and small GaAs footprint, made it the preferred option for this application. The receiver preamplifier schematic is given in Figure 10.10. A 7-turn spiral inductor ($L \approx 12.6\,\text{nH}$) was used as the tuning element. The receiver circuit comprises three amplification stages, each one using a $5{\times}150\,\mu\text{m}$ MESFET biased at $0.2\,I_{dss}$ for low-noise operation. At the front-end a low noise common source configuration is used, followed by a cascode gain stage. The output stage is common source, providing further gain and with drain load optimised to drive a $50\,\Omega$ impedance. The circuit provides an on-chip bias arrangement for the photodiode. AC coupling is used between the photodiode and the first stage input to improve the dynamic range by preventing possible bias variation due to high input average optical powers, a condition commonly encountered in SCM systems.

Over the operating range $1.8 - 2.2\,\text{GHz}$ the simulated average noise spectral density is less than $5\,\text{pA}/\sqrt{\text{Hz}}$ and the transimpedance is $72 \pm 1.5\,\text{dB}\Omega$, over the entire operating band, as shown in Figures 10.8 and 10.9 for the series tuning case. A typical pin photodiode with $0.3\,\text{pF}$ capacitance and $10\,\Omega$ bulk resistance was assumed for the simulation and optimisation results.

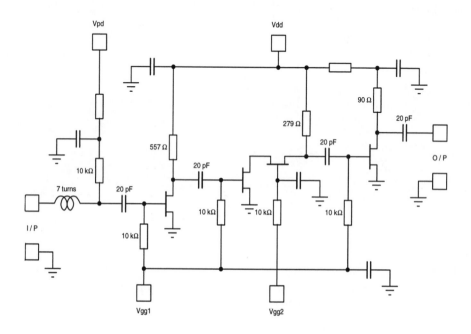

Figure 10.10 *Tuned receiver schematic*

10.5 Summary

In this chapter the basis for the design and realisation of optical fibre receivers suited to SCM applications has been considered and circuit noise modelling and optimisation discussed in detail. A relation has been established that allows the optical receiver equivalent input noise current spectral density to be calculated using the noise-matching network small signal parameters and the front-end noise parameters $F_{\min}$, Y_{opt} and R_n. From this it is possible to derive optimum noise-matching condition for an optical receiver which differs from the optimum matching for minimum noise figure, as proposed in recent studies [27].

The problem has been explored further in the context of a specific practical realisation and the question of integrating the tuning network and the preamplifier as an MMIC examined. Four tuning networks, suitable for integration, are identified and equations derived to describe the noise behaviour of optical receivers employing such networks. These equations are primarily used to obtain an initial assessment of the performance of a tuned network and to generate starting values for GaAs implementation. A receiver optimisation strategy suited to SCM applications and constrained by degeneracies associated with MMIC processes has been detailed and applied to the four types of tuning network.

Following the design and optimisation strategies developed, an integrated GaAs tuned optical receiver intended for use in a subcarrier multiplex system has been designed. The receiver is noise-optimised over the 1.6 to 2.4 GHz range, with simulation results predicting $5\,\mathrm{pA}/\sqrt{\mathrm{Hz}}$ average input noise and a transimpedance of 72 dBΩ. The on-chip input tuning arrangement employed makes the receiver easily reproducible, robust and suitable for use in practical systems.

10.6 References

1. Personick, S.D.: 'Towards Global Information Networking', *IEEE Proc.,* 1993, **18**, pp. 1549-1557
2. Mochida, Y.: 'Technologies for local access fibering', *IEEE Comms. Magazine,* 1994, **32-2**, pp. 64-73
3. Wilson, B., and Ghassemlooy, Z.: 'Pulse time modulation schemes for optical communications: a review' *IEE Proc. J,* 1993, **140**, pp. 346-357
4. Kazovsky, L., Barry, C., Hickey, M., Noronha Jr., C., and Poggiolini, P.: 'WDM local area networks', *J. Lightwave Technol. and Systems Magazine,* 1992, pp. 8-15
5. Darcie, T.E.: 'Subcarrier multiplexing for lightwave networks and video distribution systems', *IEEE J. on Selected Areas in Comms.,* 1990, **SAC-8**, pp. 1240-1248
6. Olshansky, R., Lanzisera, V.A., and Hill P.M.: 'Subcarrier multiplexed lightwave systems for broad-band distribution', *J. of Lightwave Technl.,* 1989, **LT-7**, pp. 1329-1341
7. Way, W.: 'Subcarrier multiplexed lightwave system design considerations for subscriber loop applications', *J. of Lightwave Technol.,* 1989, **LT-7**, pp. 1806-1341
8. Darcie, T.E., Dixon, M.E., Kasper, B.L., and Burrus, C.A.: 'Lightwave system using microwave multiplexing', *Electron. Lett.,*1986, **22**, pp. 774-775
9. Maeda, M., and Yamamoto, M.: 'FM-FDM optical CATV transmission experiment and system design for MUSE HDTV signals', *IEEE J. on Selected Areas in Comms ,* 1990, **SAC-8**, pp. 1257-1267
10. Fox , J.R., and Jewell S.T.: 'A broad-band distributed-star network using subcarrier fibre transmission', *IEEE J. on Selected Areas in Comms,* 1990, **SAC-8**, pp. 1223-1228
11. Van Der Ziel, A.:, 'Thermal noise in field-effect transistors' *IRE Proc.,* 1962, **50**, pp. 1808-1812
12. Van Der Ziel, A.: 'Gate noise in field effect transistors at moderately high frequencies', *IEEE Proc.,* 1963, **51**, pp. 461-467

13. Cappy, A.: 'Noise modelling and measurement techniques', *IEEE Trans. on Microwave Theory and Techniques ,*1988, **MTT-36**, pp. 1-10

14. Rohde, U.: 'Improved noise modelling of GaAs FETs, Part 1: Using an enhanced equivalent circuit technique', *Microwave J.*, 1991, **11**, pp. 87-101

15. Heymann, P. and Prinzler, H.,: 'Improved noise model for MESFETs and HEMTs in lower gigahertz frequency range', *Electron. Lett.*, 1992, **28**, pp. 611-612

16. Heinen, Kunisch, J., and Wolff, I: 'A unified framework for computer-aided noise analysis of linear and nonlinear microwave circuits', Proc. MTT-S, 1991, IEEE, NJ, pp. 1217-1220

17. Pucel, R.A., Struble, W., Hallgren R., And Rohde, U. L.: 'A general noise de-embedding procedure for packaged two-port linear active devices', *IEEE Trans. on Microwave Theory and Techniques,* 1992, **MTT-40**, pp. 2013-2024

18. Rohde, U.: 'Improved noise modelling of GaAs FETs, Part 2: Using a noise de-embedding technique', *Microwave J.*, 1991, **12**, pp. 87-101

19. Hillbrand, H., and Russer, P.: 'An efficient method for computer aided noise analysis of linear amplifier networks' *IEEE Trans. on Circuits and Sys.,*1976, **CS-23**, pp. 235-238

20. O'Reilly, J. J., and Moreira P.M.R.S.: 'Signal design for multi-Gbit/s optical receivers', Proc. Conference on information science and systems, 1992, Princeton-USA, pp. 101-105

21. Darwazeh, I., Lane, P., Marnane, W., Moreira, P., Watkins, L., Capstick, M., and O'Reilly, J.: 'GaAs MMIC optical receiver with embedded signal processing', *IEE Proc. J,* 1992, **139**, pp. 241-243

22. Hullet J.L., and Muoi, T.V.: 'A modified receiver for digital optical fiber transmission systems', *IEEE Trans. on Comms.,*1975, **COM-23**, pp. 1518-1521

23. Gimlett, J.L.: 'Low noise 8 GHz optical receiver', *Electron. Lett.,* 1987, **23**, pp. 281-283

24. Violas, M.A.R., Heatley, D.J.T., Duarte, A.M., and Beddow D.M.: '10 GHz bandwidth low-noise optical receiver using discrete commercial devices', *Electron. Lett.,*1990, **26**, pp. 34-35

25. Park, M.S., and Minasian, A.M.:, 'New low-noise tuned P-I-N-HEMT optical receiver for 10 Gbit/s signal detection', Digest of OFC/IOOC` 93 conference, 1993, San Jose-USA, pp. 15-16

26. Kimber, E.M., Patel, B.L., and Hadjifotiou, A.: '12 GHz PIN-HEMT optical receiver front end', Digest of IEE-colloquium on optical detectors and receivers, 1993, London-UK, pp. 7/1-7/10

27. Park, M.S., and Minasian, A.M.:, 'Ultralow noise 10 Gbit/s p-i-n-HEMT optical receiver', *IEEE Photon. Technol. Lett.,*1993, **5**, pp. 161-162

28. Moustakas, S. and Hullett, J.L.,: 'Noise modelling for broadband amplifier design', *IEE Proc. G,* 1981, **128**, pp. 67-76

29. Nordholt, E.H.: 'Design of high-performancence negative-feedback amplifiers', volume 7 of Studies in electrical and electronic engineering (Elsevier Scientific Publishing Company, 1983)

30. Hagen, J.B.,: 'Noise parameter transformations for three-terminal amplifiers' *IEEE Trans. Microwave Theory and Techniques*, 1990, **38**, pp. 319-321

31. Personick, S.D.: 'Receiver design for digital fibre optic communications - I', *Bell Sys. Tech. J.*, 1973, **52**, pp. 843-874

32. Personick, S.D.: 'Receiver design for digital fibre optic communications - II', *Bell Sys. Tech. J.*, 1973, **52**, pp. 875-886

33. Smith, R.G., and Personick, S.D.: 'Receiver design for optical fiber communication systems', in KRESSEL, H. (ed): Semiconductor devices for optical communication (Springer-Verlag, 1980)

34. Chausi, M.S.: Principles and design of linear active circuits (McGraw-Hill, 1965)

35. Abidi, A.A.: 'On the choice of optimum FET size in wide-band transimpedance amplifiers', *J. of Lightwave Technol.*, 1988, **LT-6**, pp. 64-66

36. Lo, D.C.W., and Forresr, S. R.: 'Performance of $In_{0.53}Ga_{0.44}As$ and InP junction field effect transistor for optoelectronic integrated circuits; Part I: Device analysis', *J. of Lightwave Technol.*, 1989, **LT-7**, pp. 957-965

37. Lo, D.C.W., and Forrest, S.R.: 'Performance of $In_{0.53}Ga_{0.44}As$ and InP junction field effect transistor for optoelectronic integrated circuits; Part II: Optical receiver analysis', *J. of Lightwave Technol.*, 1989, **LT-7**, pp. 966-971

38. Moreira, P., Darwazeh, I., and O'Reilly, J.: 'Design of an integrated tuned front-end GaAs receiver for SCM applications', Dig. of 4th Bangor communications symp., 1992, Bangor-UK, pp. 180-183

39. Moreira, P., Darwazeh, I., and O'Reilly, J.: 'Distributed amplifier signal shaping strategy for multigigabit digital optical transmission', *Electron. Lett.,*1993, **29**, pp. 655-657

40. Moreira, P., Darwazeh, I., and O'Reilly, J.:, 'Noise optimisation of tuned integrated GaAs receiver', Proc. International symp. on fibre optic networks and video comms. - SPIE-93, Vol. 1974-Transport for broadband optical access networks, 1993, Berlin-Germany, pp. 20-25

41. Jacobsen, G., Kan, J.K., and Garrett, I.: 'Tuned front-end design for heterodyne optical receivers", *J. of Lightwave Technol.*, 1989, **LT-7**, pp. 105-114

42. Darcie, T.E., Kasper, B.L., Talman, J.R., and Burrus, C.A.: 'Resonant p-i-n -FET receivers for lightwave subcarrier systems', *J. of Lightwave Technol.*, 1988, **LT-6**, pp. 582-589

43. Liu, Q.Z.: 'Unified analitical expressions for calculating resonant frequencies, transimpedances, and equivalent input noise current densities of tuned optical receiver front ends', *IEEE Trans. on Microwave Theory and Techniques,*1992, **MTT-40**, pp. 329-337

44. Jensen, N.G., Bodtker, E., Jacobsen, G., and Sorensen, S.: 'Balanced tuned receiver front end with low noise and high common mode rejection ratio', *Electron. Lett.,*1991, **27**, pp. 234-235

45. Ladbrooke, P.H.: MMIC design: GaAs FETs and HEMTS (Artecch House, 1989)

46. Moreira, P.M.R.S.: 'Optical receiver design and optimisation for multi-gigahertz applications', PhD thesis, School of Electronic Engineering Science, University College of North Wales, November 1993

47. Ha, T.T.: Solid-state microwave amplifier design (Krieger Publishing Company, 1991)
48. Medley, M.W.: 'Microwave and RF circuits: analysis, synthesis and design' (Artech House, first edition 1992)
49. Rothe, H., and Dahlke, W.: 'Theory of noisy fourpoles', *Proc. IRE*, 1956, **44**, pp. 811-818
50. Hartmann, K.: 'Noise characterization of linear circuits' *IEEE Trans. on Circuits and Sys.*, 1976, **23**, pp.581-590

Appendix: The correlation matrix noise analysis method

In this chapter the correlation matrix method proposed by Hilbrand and Russer [19] has been used. In this analysis the noisy two-port network is seen as an interconnection of more basic two-ports whose noise behaviour is known. Although more general methods are available [16-18], the types of amplifier topologies usually employed at microwave frequencies can be generally considered as two-ports. Additionally, the Hillbrand and Russer method can be simply and efficiently used for computer-aided noise analysis.

To describe a two-port network as the interconnection of more basic two-ports the following operations need to be effected [47]: series connection, parallel connection and cascade connection. If the admittance, impedance and chain matrix representations are chosen to describe the linear two-ports then, in the appropriate representation, each of these connections corresponds to a simple matrix operation. That is, if two two-port networks N_1 and N_2 are to be interconnected to form the two-port network N, then the matrix describing the resulting two-port is obtained by [47]:

$$\text{Parallel:} \quad \mathbf{Y} = \mathbf{Y}_1 + \mathbf{Y}_2 \tag{10.50}$$

$$\text{Series:} \quad \mathbf{Z} = \mathbf{Z}_1 + \mathbf{Z}_2 \tag{10.51}$$

$$\text{Cascade:} \quad \mathbf{C} = \mathbf{C}_1 \mathbf{C}_2 \tag{10.52}$$

where $\mathbf{Y}$, $\mathbf{Z}$, and $\mathbf{C}$ represent the two-port matrices in the admittance, impedance and chain representations, respectively, and the subscripts 1 and 2 refer to the two-ports to be connected[4]. The different representations can be easily converted one to another [48] so that the mathematical operation corresponding to the desired interconnection is implemented in the more convenient representation.

Additionally, any noisy linear two-port can be replaced by a noise equivalent circuit that consists of the original two-port, now assumed to be noiseless, and two additional noise sources [49,50]. This is shown in Figure 10.11 for the three representations considered.

---

[4] In the chain representation N_1 and N_2 correspond to the input and output networks, respectively.

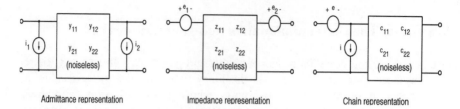

Admittance representation   Impedance representation   Chain representation

Figure 10.11 *Two-port noise models*

The noisy network is then represented by its linear two-port matrix ($\mathbf{Y}$, $\mathbf{Z}$ or $\mathbf{C}$) and its correlation matrix ($\mathbf{C}_Y$, $\mathbf{C}_Z$ or $\mathbf{C}_c$) whose elements are the power and cross-correlation spectral densities of the equivalent noise sources and are defined as:

$$\mathbf{C}_Y = \begin{bmatrix} \dfrac{d}{df}\langle i_1\, i_1^*\rangle & \dfrac{d}{df}\langle i_1\, i_2^*\rangle \\ \dfrac{d}{df}\langle i_1^*\, i_2\rangle & \dfrac{d}{df}\langle i_2\, i_2^*\rangle \end{bmatrix} \tag{10.53}$$

$$\mathbf{C}_Z = \begin{bmatrix} \dfrac{d}{df}\langle e_1\, e_1^*\rangle & \dfrac{d}{df}\langle e_1\, e_2^*\rangle \\ \dfrac{d}{df}\langle e_1^*\, e_2\rangle & \dfrac{d}{df}\langle e_2\, e_2^*\rangle \end{bmatrix} \tag{10.54}$$

$$\mathbf{C}_c = \begin{bmatrix} \dfrac{d}{df}\langle e\, e^*\rangle & \dfrac{d}{df}\langle e\, i^*\rangle \\ \dfrac{d}{df}\langle e^*\, i\rangle & \dfrac{d}{df}\langle i\, i^*\rangle \end{bmatrix} \tag{10.55}$$

If the noisy networks N_1 and N_2 are interconnected to form the two-port network N, their correlation matrices must be combined as [19]:

Parallel: $\mathbf{C}_Y = \mathbf{C}_{Y1} + \mathbf{C}_{Y2}$ \hfill (10.56)

Series: $\mathbf{C}_Z = \mathbf{C}_{Z1} + \mathbf{C}_{Z2}$ \hfill (10.57)

Cascade: $\mathbf{C}_c = \mathbf{C}_{c1} + \mathbf{C}_1\, \mathbf{C}_{c2}\, \mathbf{C}_1^+$ \hfill (10.58)

where $^+$ denotes Hermitian conjugation.

Coversion between representations for the correlation matrices are given by the transformation

$$\mathbf{C}_R = \mathbf{T}\,\mathbf{C}_O\,\mathbf{T}^+ \tag{10.59}$$

where $\mathbf{C}_O$ and $\mathbf{C}_R$ denote the correlation matrix of the original and resulting representation, respectively, and $\mathbf{T}$ is the transformation matrix between the two representations. This matrix is given in Table 10.2 for transformations between the admittance, impedance and chain representations [19].

Finally, for a reciprocal circuit consisting only of passive elements the correlation matrix can be obtained in the admittance or impedance representations as:

$$\mathbf{C}_Y = 2\,k\,\theta\,\mathrm{Re}(\mathbf{Y}) \tag{10.60}$$

$$\mathbf{C}_Z = 2\,k\,\theta\,\mathrm{Re}(\mathbf{Z}) \tag{10.61}$$

Table 10.2 *Transformation matrices*

Resulting representation	Original representation		
	Y	**Z**	**C**
Y	$\begin{bmatrix} 1 & 0 \\ 0 & 1 \end{bmatrix}$	$\begin{bmatrix} y_{11} & y_{12} \\ y_{21} & y_{22} \end{bmatrix}$	$\begin{bmatrix} -y_{11} & 1 \\ -y_{21} & 0 \end{bmatrix}$
Z	$\begin{bmatrix} z_{11} & z_{12} \\ z_{21} & z_{22} \end{bmatrix}$	$\begin{bmatrix} 1 & 0 \\ 0 & 1 \end{bmatrix}$	$\begin{bmatrix} 1 & -z_{11} \\ 0 & -z_{21} \end{bmatrix}$
C	$\begin{bmatrix} 0 & c_{12} \\ 1 & c_{22} \end{bmatrix}$	$\begin{bmatrix} 1 & -c_{11} \\ 0 & c_{21} \end{bmatrix}$	$\begin{bmatrix} 1 & 0 \\ 0 & 1 \end{bmatrix}$

Chapter 11

Optical fibre amplifiers and WDM

M. J. O' Mahony

11.1 Introduction

Optical amplifiers have become a major element in the design of advanced optical systems. Their main characteristic is that they provide direct optical gain over a very wide bandwidth. In addition, they can support very high output powers and so are ideal for a number of analogue applications, such as CATV distribution. The wide bandwidth enables the simultaneous amplification of many signals at different wavelengths, making them ideal components to use in wavelength division multiplexed systems.

11.2 Optical amplifiers

Optical amplifiers allow the direct amplification of light, without the need for optical to electrical conversion [1,2]. They provide the amplification necessary to overcome system losses associated with fibre and other optical components. Fibre amplifiers can support very high output powers (>100 mW) and are ideal for optical distribution systems.

The importance of the optical amplifier is that it provides high gain (up to 50 dB) over extremely wide bandwidths (approximately 3.7 THz); this performance is unequalled by any electronic component. The wide bandwidth of the amplifier matches reasonably well that available from the fibre itself. An optical system using amplifiers is therefore said to be transparent in that the end-to-end bandwidth is so great that it can accommodate signals of any format or bit-rate. This notion of transparency is viewed as an important attribute of future systems as it provides flexibility for system operators. In principle a system using optical amplifier repeaters can support different bit rates, without the need for changing hardware as required with opto-electronic regenerators. Such a system offers the potential of allowing capacity upgrading with hardware changes necessary only at the terminals. Optical amplifiers can also support the simultaneous amplification of a number of channels on different wavelengths. Thus a system using optical amplifiers can accommodate a mixture of analogue and digital services on different wavelengths.

Optical amplifiers are now being considered for a wide range of applications and are becoming an essential element in many future system and network designs. In point-to-point systems they can be used as power boosters, repeaters or receiver preamplifiers [3]. In optical networks they can be used as functional devices to perform wavelength conversion in wavelength division multiplexed systems (i.e. changing the carrier wavelength of a specific information stream) or as gates to form optical switch matrices [4]. In distribution networks they can be used as high power gain blocks to increase the maximum splitting ratio [5].

This discussion is confined to amplifiers based on optical fibre, and in particular erbium doped fibre amplifiers (EDFAs). Although many other types of amplifiers based on fibre or semiconductor material are available, their characteristics are not as suitable as those of EDFAs for analogue applications. Amplifiers operate through the process of stimulated emission (and the attendant processes of spontaneous emission and absorption) where signal photons incident on the amplifier active material cause the generation of additional photons which are of the same frequency, polarisation and phase as the incident photons. Stimulated emission requires the condition of population inversion in the active material, in which the population of material energy states is changed from its normal distribution where the lower energy states are most heavily populated. In a condition of population inversion the higher energy states are more heavily populated through optical pumping of the erbium doped fibre. The degree of population inversion determines the available amplifier gain and may be increased by increasing the optical pump power.

In any real application of optical amplifiers a number of parameters are required to establish performance; the major ones are:

- Net gain: the gain from input to output fibre
- Bandwidth: the bandwidth associated with the gain process
- Noise: the amount of noise generated by the amplification process
- Polarisation sensitivity:
             the sensitivity of device gain to variations in polarisation of the incident signal
- Output saturation power:
             the output signal power at which the gain has fallen to 3 dB of its unsaturated value
- Linearity: linearity associated with the gain transfer characteristic

The following section describes the basic operation of an EDFA and its main characteristics.

11.3 Rare-earth doped-fibre amplifiers

Amplifiers based on silica fibres doped with rare earths such as erbium, praseodymium or ytterbium have a number of intrinsic advantages. In particular, the amplifying medium is easily spliced to the system fibre with little loss and the amplifier is polarisation insensitive. Figure 11.1 is a schematic diagram of an EDFA which comprises a length of amplifying fibre (silica fibre doped with erbium, whose length is generally less than 50 m), an optical pump to provide the energy necessary for population inversion and a coupler to combine signal and pump in the amplifying fibre. In practice the pump power can be injected in a number of ways. For example, it may be injected so that it propagates in the same direction as the signal, termed co-directional pumping. Alternatively the pump can be injected at the far end (counter-directional pumping), or indeed for high output power amplifiers the fibre may be pumped at both ends, as illustrated. Co-directional pumping is generally employed to give the best noise performance. Because of the very high gains (50 dB) that can be supported by a fibre amplifier, many commercial amplifiers include an isolator (sometimes two) at the fibre ends to eliminate reflections back into the fibre which could cause instability.

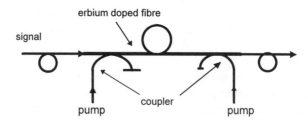

Figure 11.1 *Schematic diagram of erbium-doped fibre amplifier*

The passband characteristic of an erbium amplifier depends critically on the co-dopants (generally alumina and germania) as well as the erbium, rather than on the silica material, which plays the part of a host medium. Other rare-earth dopants may be used to provide amplification at other wavelengths; for example, amplifiers designed to operate at 1.3 μm use praseodymium [6]. Figure 11.2 shows the energy levels associated with an erbium atom.

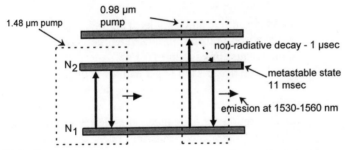

Figure 11.2 *Energy levels*

To achieve population inversion it is necessary to optically pump the amplifying fibre at wavelengths corresponding to the energy level transitions shown in the diagram. The pump wavelengths most commonly used are 1.48 µm and 0.98 µm. At 1.48 µm, the absorption of a pump photon excites an electron directly from the ground state to the metastable state. A photon is then emitted when a transition occurs between the metastable and the ground states caused by either random spontaneous decay or by emission stimulated by an incident photon. A pump photon at 980 nm causes excitation to a higher energy state, from which level the electron falls back to the metastable state (releasing a certain amount of energy in the process). Once again transitions from the metastable to ground states are accompanied by photon emission.

11.3.1 Rate equations

Evaluation of the overall gain characteristic requires solutions to the propagation equation associated with the signal and pump travelling along the fibre, and the rate equations which define the transitions between energy states of the erbium atom [7]. The rate equations describe the variation of the population densities, i.e. the number of erbium atoms per volume element in each energy state. The transition rate associated with an energy state is defined as the population density times the rate coefficient.

It is convenient to take the example of the two-level 1.48 µm pumped amplifier to discuss the main features of amplifier operation, although more accurate and detailed modelling may be found in a number of publications, e.g. Reference 8. The rate equation defining the population densities in the two-level system can be expressed in the form (for the upper level):

$$\frac{dN_2}{dt} = (R_{12} + W_{12})N_1 - (R_{21} + W_{21} + A_{21})N_2 \tag{11.1}$$

where

R_{12}, R_{21}	= pump absorption and stimulated emission rates
$W_{12} \; W_{21}$	= signal absorption and stimulated emission rates
A_{12}	= spontaneous emission rate ($= 1/\tau$)
N_1, N_2	= population densities of upper and lower energy states

The spontaneous lifetime τ of the excited state is approximately 11 ms. Assuming the pump absorption and stimulated emission rates are equal (W_p) and the signal absorption and stimulated emission rates are equal (W_s) and no stimulated pump emission occurs, then the rate equations for the upper and lower levels are [9]:

$$\frac{dN_2}{dt} = W_p N_1 - W_s(N_2 - N_1) - \frac{N_2}{\tau} \qquad \text{upper level}$$

$$(11.2)$$

$$\frac{dN_1}{dt} = -\frac{dN_2}{dt} \qquad \text{lower level}$$

and the total density is $N_t = N_1 + N_2$. Expressing the transition rates in terms of powers then gives:

$$W_p = \frac{P_p \sigma_{pa}}{A_p h \nu_p}; \qquad W_s = \frac{P_s \sigma_{se}}{A_s h \nu_s} \qquad (11.3)$$

where $\sigma_{pa/se}$ are the pump absorption and signal emission cross-sections, respectively, $A_{s/p}$ the cross-sectional areas of the signal and pump modes within the fibre core and $P_{s/p}$ the signal and pump powers.

A cross-section is an effective area presented to an incident field by an ensemble of emitting or absorbing ions and in general the absorption and emission cross-sections are not equal. Equation 11.2 may be solved for the steady state ($d/dt=0$) to give [9]:

$$N_2 = \frac{(P_p' + P_s')N_t}{1 + 2P_s' + P_p'} \qquad\qquad N_1 = \frac{(1 + P_s')N_t}{1 + 2P_s' + P_p'} \qquad (11.4)$$

where $P_p' = P_p/P_p^{sat}$ and $P_s' = P_s/P_s^{sat}$ and:

$$P_p^{sat} = \frac{A_p h \nu_p}{\sigma_{pa} \tau}; \qquad\qquad P_s^{sat} = \frac{A_s h \nu_s}{\sigma_{se} \tau} \qquad (11.5)$$

Gain saturation occurs when the injected signal perturbs the population of the upper level to such an extent that the difference between the energy levels is appreciably diminished. A measure of this is when the factor N_2-N_1 has decreased to half its unsaturated value. It can be shown from Equations 11.2-11.4 that this occurs when the signal power equals P_s^{sat} as defined in Equation. 11.5.

11.3.2 Propagation equations

The rate equations describe the behaviour of the population densities for a given pump and signal power P_p and P_s, with A and σ as material parameters. The pump and signal powers vary along the amplifier length because of absorption, stimulated emission and spontaneous emission, thus the population densities N_1 and N_2 are functions of z. The gain coefficient at a point z along the fibre may be defined as:

$$g(z) = N_2(z)\sigma_{se}(v) - N_1(z)\sigma_{sa}(v) \tag{11.6}$$

Ignoring the contribution of spontaneous emission, the forward propagation equations relating to pump and signal are:

$$\frac{dP_s(z)}{dz} = g(z)\, P_s(z) \tag{11.7}$$

$$\frac{dP_p(z)}{dz} = -N_1(z)\sigma_{pa}\, P_p(z) \tag{11.8}$$

Thus the gain coefficient depends on the population inversion, which in turn depends on the pump power. As the pump power level decreases along the fibre length, therefore, so does the signal gain (see Figure 11.5). Using these and the earlier equations one can obtain approximate expressions for the variation of signal power with fibre length [9]:

$$\frac{dP_s}{dz} = \frac{\alpha_s P_s(P_p' - 1)}{1 + 2P_s' + P_p'}$$

$$\frac{dP_p}{dz} = -\frac{\alpha_p P_p(1 + P_s')}{1 + 2P_s' + P_p'} \tag{11.9}$$

where $\alpha_s = \sigma_{sa}N_t$, $\alpha_p = \sigma_{pa}N_t$ (in which $N_t = N_2 + N_1$) represent the absorption coefficients at the signal and pump frequencies, respectively; with the simplifying assumption that they are equal.

Within the fibre core the erbium atoms have a particular distribution. Accurate modelling requires that the gain coefficient takes account of this distribution. Therefore, in general the overall gain coefficient at z is found by integrating over the fibre cross-section, such that:

$$g(z,\lambda) \quad = 2\pi \int \rho(r)\psi(r)\big[\sigma_{se}N_2(r,z)-\sigma_{sa}N_1(r,z)\big] \, rdr \qquad (11.10)$$

where $\rho(r)$ is the erbium distribution function and $\psi(r)$ is the optical beam profile (which, for example, can be approximated by a Gaussian shape).

The simultaneous solution of the rate and propagation equations defines the amplifier behaviour. Accurate analysis requires a computational approach.

11.3.3 Fibre amplifier characteristics

Figure 11.3 is an example of a measured overall gain characteristic for a 20 m amplifier with a pump power of 20 mW [10], showing the gain peak at 1.53 µm, characteristic of an erbium amplifier. It also illustrates that erbium amplifers display bandwidths in the order of 30 nm, but that the pass band is not completely flat. This latter feature is problematic in systems using cascades of optical amplifiers supporting wavelength division multiplex transmission, where the end-to-end passband shape will be significantly different from that of a single amplifier.

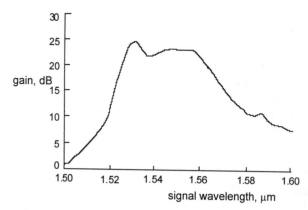

Figure 11.3 *Gain characteristic of EDFA*

The irregular pass band shape may also cause effects which are problematic in analogue applications. The combination of fluctuations in the spectrum of the input signal, due to laser chirp for example, and the slope of the gain response will cause frequency to amplitude conversion, which appears as a source of noise.

The choice of an operating point, at which the slope of the characteristic is small, may be necessary to minimise the effect of this impairment.

Equations 11.1-11.9 show that the amplifier gain is a complicated function of pump power and amplifier length. For a given pump power, for example, the gain reaches a maximum for a particular fibre length and then decreases. This behaviour is illustrated schematically in Figure 11.4.

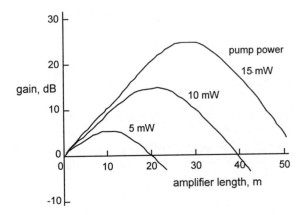

Figure 11.4 *EDFA gain against fibre length*

Thus in designing an amplifier the fibre length is chosen to match the pump power in order to achieve a particular gain.

The saturation power of the amplifier is given by Equation. 11.5, namely:

$$P_s^{sat} = \frac{A_s h v_s}{\sigma_{se} \tau}$$

where A_s is the cross-sectional area of the fibre core. A typical value of this parameter is about 5 mW, however, as discussed below, in the case of erbium fibre amplifiers saturation power is not equivalent to the maximum operating power.

The discussion above has centred on the erbium fibre amplifier. Unfortunately, the significant disadvantage of the EDFA is that it operates only in the 1.5 μm region. In the UK, the vast majority of optical systems operate at 1.3 μm, precluding the upgrading of such systems by EDFAs. For this reason research has intensified into amplifiers which operate at 1.3 μm. Figure 11.5 is an example of the passband characteristic of a 1.3 μm amplifier based on the rare-earth dopant praseodymium.

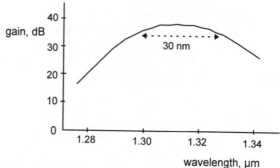

Figure 11.5 *Gain characteristic of 1.3 μm amplifier*

It can be seen that the passband of the praseodymium amplifier is more symmetrical than that of the EDFA, with a similar gain bandwidth of approximately 30 nm. The main disadvantage of these amplifiers is that the required pump power is sufficiently high to preclude the use of semiconductor pumps, thus increasing the cost.

11.3.4 Amplifier noise

In an amplifier exhibiting population inversion there is a natural decay or transition rate from upper to lower energy states. Such transitions result in the emission of photons unrelated to the signal, and this spontaneous emission is the source of noise within amplifiers as it adds to the signal during amplification.

To analyse the noise contributions consider an optical amplifier of gain G positioned in front of a photodetector with responsivity R Amps/Watt. Assuming the optical amplifier input power is P_{in} then the mean photodetector current I_o is:

$$\langle I_o \rangle = GRP_{in} + RP_{ase}; \quad P_{ase} = 2\gamma(G-1)h\nu \,\Delta f \qquad (11.11)$$

where Δf is the amplifier bandwidth, γ is the population inversion parameter, which for the two-level amplifier has the form $\gamma = N_2/[N_2-(\sigma_{sa}/\sigma_{se})N_1]$. As discussed above, both N_2 and N_1 will vary along the amplifier length in both fibre and semiconductor amplifiers and so the analysis here is of an approximate nature. The first term in Equation 11.11 represents the amplified signal and the second term the amplified spontaneous emission (ASE). The latter is particularly important in cascaded amplifier systems, as the ASE accumulates in direct proportion to the number of amplifiers.

To establish the signal-to-noise ratio, the variance of the output noise must be evaluated [11]. This has four terms, namely:

$$\langle i^2_{sig.shot} \rangle \;=\; 2e(P_{in}GR)B$$

$$\langle i^2_{sp.shot} \rangle \;=\; 2eRP_{ase}B$$

$$\langle i^2_{sig.sp} \rangle \;=\; 4eP_{in}GR\gamma\,(G-1)B$$

$$\langle i^2_{sp.sp} \rangle \;=\; 2eR\gamma\,(G-1)P_{ase}B \qquad (11.12)$$

where e is the electron charge and B the electrical bandwidth. The first two terms represent the shot noise associated with the signal and amplified spontaneous emission. The last two terms represent the beat noise associated with interaction between the signal and spontaneous photons and between spontaneous photons respectively. In practice an optical filter may be used to significantly reduce the magnitude of the spontaneous-spontaneous beat noise and the spontaneous shot noise, and in this case the dominant noise components are the signal-spontaneous beat noise and signal shot noise. The signal-to-noise ratio is then:

$$SNR \;=\; \frac{(P_{in}GR)^2}{2e(P_{in}GR)B+4eP_{in}GR\gamma\,(G-1)B} \;\cong\; \frac{GP_{in}R}{2eB(1+2\gamma\,(G-1))} \qquad (11.13)$$

The concept of a noise figure is often applied to optical amplifiers. Assuming that all but the signal-spontaneous beat noise components are, or can be made, negligible then the noise figure F_n is the ratio of the quantum limited *SNR* at the input to the output *SNR*, i.e.:

$$F_n \;=\; \frac{(SNR)_{in}}{(SNR)_{out}} \;=\; \frac{1+2\gamma\,(G-1)}{G} \qquad (11.14)$$

This equation shows that the *SNR* is degraded by a factor of 2 (3 dB) even for an ideal amplifier in which $\gamma=1$ and $G \gg 1$. For fibre amplifiers pumped at 980 nm noise figures very close to 3 dB are obtainable (e.g. 3.2 dB). For amplifiers pumped at 1.48 μm the noise performance is generally worse.

11.4 Wavelength division multiplexing

The main transmission windows for telecommunications are centred around 1.3 μm and 1.5 μm [12]. Currently 1.3 μm is the wavelength most commonly used, but the high-performance long-distance systems generally use 1.5 μm because of its lower loss. The 1.5 μm window has also gained particular importance because of the availability of EDFAs. Both windows of operation have bandwidths in the order of 100 nm, or approximately 12 THz. In recent years technology developments have made it possible to make more efficient use of this available bandwidth by the simultaneous transmission of information on different wavelengths, a technique generally known as wavelength division multiplexing (WDM) [13]. For example, in an all-optical access network, telephony services could be carried on one wavelength, and a variety of broadband services (video etc.) on an additional set of wavelengths [14]. The performance of wavelength multiplexing is considerably enhanced by the ability of fibre optical amplifiers to provide simultaneous amplification of multiple channels. Thus wavelength division multiplexing and optical amplification are essential elements for many future infrastructures supporting both analogue and digital signals.

The concept of WDM is illustrated schematically in Figure 11.6. The diagram shows how four signals on different wavelengths are combined by a multiplexer on to a single fibre. At the far end a demultiplexer is used to separate out the individual wavelengths.

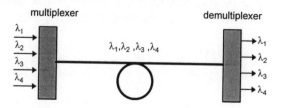

Figure 11.6 *Wavelength division multiplexing*

WDM is a technique which is important for both point-to-point systems, all-optical networks and optical distribution networks. In point-to-point systems it can be used to improve fibre utilisation, as illustrated in Figure 11.6, and this can be important in situations where, for example, duct space is at a premium. Additionally, it can be used to overcome the transmission limitations in digital systems due to dispersion, by dividing the total required capacity between a number of wavelengths; for example, 4 x 2.5 Gbit/s can be used to realise a capacity of 10 Gbit/s. In optical networks [15] multiplexing is used as a means of providing multiple-access; that is a large number of individual connections can be established over a common fibre with wavelength defining the interconnection path [16]. In optical distribution networks WDM can be used to support both analogue and digital services.

The main advantages of WDM are that (a) many signals can be carried on one fibre, thus capacity is increased with little additional complexity, and (b) a mixture of services can be supported using different wavelengths on the same infrastructure. The main disadvantage is that new components, multiplexers and demultiplexers, are required which introduce loss and decrease the available power budget.

11.4.1 *Maximum number of wavelengths*

Although in excess of 100 nm of bandwidth is available in the 1.5 μm window, the assumption that most advanced systems will use optical amplification means that the maximum usable bandwidth is approximately 30 nm, the bandwidth of a single EDFA. The equivalent frequency range is 3.7 THz. Assuming that (a) each analogue baseband channel requires a bandwidth B of 1 GHz; (b) single sideband modulation is possible; (c) a guard-band equal to three times the bandwidth is used to allow for imprecise demultiplexing etc., then the maximum number of channels possible is $N_{max} = \Delta f / 4 B = 860$. Thus, in theory, less than 1000 channels can be accommodated within the bandwidth of a single amplifier. This number is still ideal, however, as such close packing density demands very high performance from the optical components. To avoid crosstalk between channels with this separation, extreme laser source wavelength stability and very high selectivity at the receiver are required. Any chirping of the laser source would also spread spectral components into adjacent channels. Thus the above bound is very ideal, but it is used to make the point that the number of available wavelengths is in the hundreds rather than the thousands.

As a certain power per channel is required to ensure an adequate system error rate, the total power in the fibre (the sum of the individual powers) can be very high. For example, with 1 mW/ channel, 860 channels would mean a total power close to 1 W. Such high powers expose the fundamental non-linear nature of optical fibre, which gives rise to distortion and crosstalk. Non-linear effects [17,18,19] are a function of the total power in the fibre and the interaction length, effectively constraining the maximum number of wavelengths possible over a particular distance.

11.4.2 *Classification of multi-wavelength systems*

Within the broad range covered by the general title of multi-wavelength systems, a number of specific systems are often identified. Detector selectivity is important in determining the channel wavelength separation and is a function of the available technology. For example, there is an identifiable divide between the technology required to separate channels with spacings less than or greater than approximately 1 nm (130 GHz). Well below 1 nm special techniques such as coherent detection may have to used to provide the required selectivity. Coherent detection requires the use of a laser at the receiver as well as at the transmitter in a manner analogous to the heterodyne radio, making it complex and expensive. Above 1 nm commercial optical components and filters are readily available. It is common therefore to classify systems according to the wavelength separation used. Examples are:

WDM: Systems with wavelength spacing > 1 nm (e.g. 4 nm) using direct detection techniques. Coarse optical filtering can be used prior to detection to select wavelength. These typically use a small number of channels (2-8).

HDWDM: High density WDM systems with spacing ≤ 1 nm. These generally use direct detection together with predetection narrowband optical filtering for wavelength selection.

CMC: Coherent multichannel systems with spacing << 1 nm. These use coherent heterodyne detection techniques to provide very high selectivity together with good detection sensitivity, permitting very small channel separation in the region of 0.1 nm (13 GHz) at 1.5 µm. This technique would allow in excess of 100 channels, representing the best technique for utilising the fibre bandwidth - but the most complex.

The majority of near-term applications for multi-wavelength systems use WDM. From the reasoning given above it is unlikely that we will see systems operating with more than 100 wavelengths in the foreseeable future. The following discussion therefore is confined to WDM systems and technology.

11.4.3 Components for WDM systems

Figure 11.6 showed some of the basic wavelength selective elements necessary to realise a WDM system. The main ones are optical filters, multiplexers and demultiplexers.

11.4.3.1 Optical filters
An optical filter is a basic building block which provides the necessary wavelength selection. The transmission characteristics of the filter are most important as they have a significant effect on the performance of the system as a whole. Thus, for example, the filter passband characteristic and its roll-off slope must be considered in relation to the wavelength separation to ensure minimum crosstalk. In addition, for a system using more than one optical filter, the overall passband characteristic will be different from that of an individual filter and determined by the cascade of individual characteristics, making the end-to-end bandwidth considerably less than that of an individual filter. Three types of filter are commonly used for WDM: fixed interference filters, tuneable Fabry-Perot filters and grating based filters.

Fixed interference filters. Interference filters are constructed using multiple thin-film layers. The transmission characteristic of such a filter is dependent on the specific design (number of layers etc.), but a typical three-stage double half-wave device will display a passband characteristic of the form:

$$T(\Lambda) \quad = \frac{1}{1+(2\Lambda)^6} \qquad (11.15)$$

where Λ is the normalised frequency defined as $\Lambda = (\lambda - \lambda_0)/\delta\lambda$, in which λ_0 is the centre wavelength and $\delta\lambda$ is the 3 dB bandwidth of the filter. A characteristic of this type of filter is that its response falls off very rapidly outside the passband, hence it is a very good choice for WDM. For example, for $\Lambda=2$, adjacent channel interference is -36 dB. Its tuning capability is limited, however, although it may be tuned slightly by tilting the filter with respect to the incident signal. Typical insertion loss is generally > 2 dB and bandwidths are generally > 1nm.

Tuneable Fabry-Perot filters. Figure 11.7 is a diagram of a fibre Fabry-Perot filter, a tuneable filter that has proved of great interest due to its simple construction and ready availability [20]. In the figure a cavity is shown between the ends of two fibre tails, where the fibre ends are coated to give a very high reflectivity. To tune the device one of the fibres can be moved to change the optical cavity length.

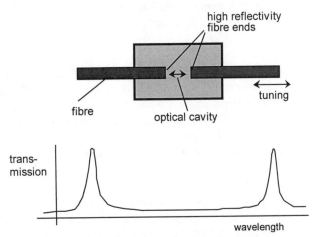

Figure 11.7 *Fibre Fabry-Perot filter and transmission characteristic*

The wavelength difference between the repetitive peaks of the resulting transmission characteristic is termed the free spectral range (*FSR*) and is a function of the cavity length. The bandwidth of the transmission mode is *FSR/F*, where F is the finesse which is determined by the reflectivity of the fibre ends. The transmission characteristic is given by :

$$T(\Lambda) \;\; = \frac{1}{1 + (\frac{2F}{\pi})^2 \sin^2(\frac{\Lambda}{F})} \;\; ; \qquad F = \; finesse = \;\; \frac{\pi \sqrt{R}}{1 - R} \tag{11.16}$$

where R is the reflectivity of the fibre end. The free spectral range (*FSR*) is given by the equation $FSR = c /(2nL_c)$, where L_c is the cavity length, n the refractive index and c the velocity of light. By way of example, typical parameters might be $L_c = 26\mu m$ and $F = 100$. A typical component has a bandwidth of 0.8 nm and an insertion loss of around 2dB. Unfortunately, the Fabry-Perot filter is not very selective. With $F = 100$ and for $\Lambda=2$, for example, adjacent channel interference is only -10 dB, making these filters suitable only for very coarsely spaced wavelength systems.

Grating based filters. Optical filters based on the use of diffraction gratings are also commonly used. The transmission characteristic can be approximated by the Gaussian function:

$$T(\Lambda) \;\; = \exp\left[-(4\ln 2)\Lambda^2\right] \tag{11.17}$$

For $\Lambda=2$, for example, adjacent channel interference is -48 dB, making grating based filters very suitable for wavelength selection. An example of a grating filter is illustrated in the discussion on demultiplexers below.

11.4.3.2 Multiplexers and demultiplexers
Multiplexing can be achieved in two ways: by a simple passive combiner or a wavelength selective combiner. Demultiplexing can only be achieved by using wavelength selective components. In the passive combiner no wavelength selection is involved, signals are simply combined, for example, through the use of fibre couplers. As a 3 dB power loss is incurred for each passive split (or combination) a serious disadvantage is that for a large number (N) of channels, the loss ($10\log_{10}N$) is very high. For example, with $N =10$, the loss at the combiner would be 10 dB; for $N = 100$ the loss is 100 dB. In practice an excess loss would also be associated with the passive combiner/splitter due to non-ideal realisation. Because of the need to select a wavelength at the receiver, a passive splitter must be used in conjunction with an optical filter, or some other wavelength selective component. The same splitter loss is incurred as with the simple combiner, but in addition the filter will also have an insertion loss of around 3 dB, for example.

Thus a WDM system using 4 wavelengths and a lossless passive combiner together with a passive splitter/filter demultiplexer would have its power budget per channel reduced by 15 dB. This illustrates one of the disadvantages mentioned earlier, namely the reduction in power budget associated with the introduction of the additional components necessary to support multiplexing and demultiplexing.

The grating multiplexer/ demultiplexer is illustrated schematically in Figure 11.8. It comprises an array of fibres, a lens and an optical grating .

In operation the multiplexed signal enters one fibre, where it is focused by the lens onto the grating. The grating returns the incident light in a beam in which the different wavelength components are spatially separated. The geometry of the device is such that the individual spatially separated components illuminate a particular fibre; thus the output fibre array comprises signals corresponding to the individual wavelengths.

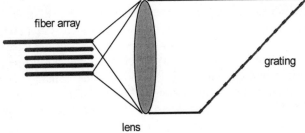

Figure 11.8 *The grating multiplexer/demultiplexer*

An important advantage of the grating demultiplexer is that the channel loss is relatively independent of wavelength number. Commercial devices with a wavelength separation of 4 nm have an insertion loss of approximately 5 dB. Grating based multiplexers are therefore advantageous for systems using large numbers of wavelengths (e.g >5), but are expensive as a result of their complicated construction.

11.5 WDM systems using optical amplifiers

The preceding discussion has focused on the technology required to implement WDM and the important issues that must be considered in the design of systems using optical amplification. It is likely that many future networks will use a combination of WDM and optical amplification and a number of important issues must be considered. Crosstalk within the optical fibre amplifier itself is likely to be negligible. However, the high output powers possible through the use of amplifiers may lead to non-linear crosstalk mechanisms within the fibre, even over relatively short distances. Linear crosstalk may arise from non-ideal wavelength selective devices.

Amplified WDM systems have similar characteristics to single wavelength amplified systems with respect to signal and noise performance. Figure 11.9 is a schematic diagram of a typical system of total length L with amplifier spacing L_A, such that the gain G of each amplifier exactly compensates for the fibre section loss.

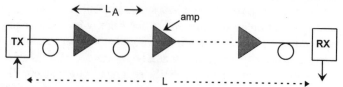

Figure 11.9 *System with optical amplifier repeaters*

As amplifiers are analogue devices, noise generated within each amplifier will accumulate along the chain of amplifiers in a cascaded system, ultimately limiting the number of amplifiers that may be cascaded. Accumulated amplifier noise limits performance in two ways (a) the signal-to-noise ratio degrades along the system length and (b) the amplified spontaneous emission (ASE) may cause saturation of amplifiers further down the chain, limiting the gain available for the signal. The latter point is significant for WDM systems as the total signal power (the sum of all the channel powers) may be quite large, resulting in the limitation imposed by accumulated ASE becoming apparent at an earlier stage in the chain than would be the case for a single wavelength system, thus restricting WDM systems to fewer cascaded amplifiers. Using Figure 11.9 as an example, and assuming each amplifier has a gain of 30 dB, which exactly matches the fibre section loss, then the total ASE power at the output of the Nth amplifier is NP_{ase}, where P_{ase} is defined by Equation 11.11. Assuming the population inversion parameter $\gamma = 1$ and a bandwidth of 30 nm, then the total ASE power equals the amplifier saturation power of 5 mW (from the earlier example) after only 10 amplifiers. It is for this reason that optical filters are often included in amplifier systems, as a narrowband filter will significantly reduce the ASE power.

In addition to noise related effects, amplified WDM systems have a number of significant differences from single wavelength systems which must be considered in system design. Some of the major ones are described below.

11.5.1 Gain flatness

The transmission characteristic of a basic erbium amplifier was illustrated in Figure 11.3. The passband illustrated in the diagram is characterised by a gain peak at a wavelength of approximately 1.53 μm with a deviation in gain across the band 1.53-1.56 μm of approximately 3 dB. For a multi-wavelength system this gain variation is of much greater significance than in a single wavelength system. For example, with a cascade of 10 amplifiers, each with a transmission characteristic as shown by Figure 11.3, it is possible that the intensity of a signal at one wavelength at the end of the cascade may be 30 dB less than another at a different wavelength - which is clearly not satisfactory. Hence, for WDM systems, it may be necessary to use gain-flattened amplifiers. Such amplifiers are constructed using an integral filter to reduce the gain peak and flatten the characteristic. However, even this approach is not straightforward, as in general the flattened characteristic is only appropriate for a particular input power and gain combination. Even with gain-flattened amplifiers, therefore, the end-to-end bandwidth after a cascade of 50 can have decreased from 30 nm to less than 10 nm. Thus, in the design of an optically amplified WDM system, it is the end-to-end bandwidth that must be considered when deciding on the number of wavelengths and the wavelength separation.

11.5.2 Crosstalk

A major source of linear crosstalk is the non-ideal filtering associated with the demultiplexing process [21]. As optical filters do not have ideal stop bands, adjacent channel interference will always be present. For a particular system design it is necessary to calculate the amount of crosstalk that can be tolerated to achieve a given signal-to-noise ratio and use this criterion to select the appropriate filter type. For example, if the adjacent wavelength rejection is to be greater than 20 dB, then a Fabry-Perot filter can only be used if the channel separation normalised by the filter bandwidth exceeds 5; i.e. for a 1 nm filter bandwidth the channel separation must exceed 5 nm.

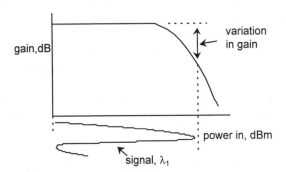

Figure 11.10 *Crosstalk in a WDM system*

A number of crosstalk mechanisms also exist within optical amplifiers. A common source in all amplifiers is that due to cross-saturation. Figure 11.10 illustrates the case where two signals are being amplified simultaneously in a device operating in the saturation regime. Assuming that signal λ_1 on its own has sufficient power to force operation into the saturation regime, then the gain of the amplifier will be modulated by the intensity variations of the signal. Any additional channel such as λ_2 will then be modulated by the varying gain, resulting in potential cross talk.

The carrier lifetime associated with the amplifier excited state is of crucial importance in this scenario, as it defines how rapidly the gain of the amplifier can respond. In the EDFA the carrier lifetime is approximately 11 ms, thus crosstalk can only occur on a very slow time constant, effectively confining crosstalk energy to frequencies in the order of kHz. This value is not generally a problem for typical telecommunication signal bandwidths. Crosstalk may also occur due to non-linear effects in the system fibre, particularly when amplifiers are operating at high output powers, as discussed below.

11.5.3 Non-linear limitations

The total power in an amplified system fibre when a large number of wavelengths are present can exceed 100 mW. In addition, for some applications, the distances involved may be large. Consequently, non-linear effects, which are a function of both total power and distance, may be observed at very low individual channel power levels. Fibre non-linearities effectively constrain the number of wavelengths x distance product, since beyond a certain value effects such as cross talk dominate performance. The main non-linearities discussed here are stimulated Raman scattering and four wave mixing, although for a number of applications cross-phase modulation may also be significant. This effect occurs when a signal at one wavelength causes a fluctuation in fibre refractive index which in turn causes a fluctuation in phase of all the other channels.

Stimulated Raman scattering. Stimulated Raman scattering (SRS) is a non-linear effect in which light in the fibre interacts with molecular vibrations and scattered light is generated at a wavelength longer than that of the incident light. If another signal is present at this longer wavelength then it will be amplified, effectively transferring energy from one signal to another. This mechanism results in crosstalk, degrading the performance of a WDM system [18]. For fused silica fibre the peak gain coefficient associated with the process is 9.4×10^{-14} m/W at a pump wavelength of 1 μm. The process can be represented by an approximately triangular gain profile [17] between 0 and 13 THz (approximately 100 nm at 1.5 μm).

$$g_r = \frac{g_r(\max)\Delta f_i}{2.6 \times 10^{13}} \qquad \text{for} \quad \Delta f_i \le 1.3 \times 10^{13}\,\text{Hz, and}$$

$$g_r = 0 \qquad\qquad \text{for} \quad \Delta f_i > 1.3 \times 10^{13}\,\text{Hz}$$

(11.18)

where Δf_i is the frequency spacing between each individual channel and the shorter wavelength, which is always placed at 0 THz. In amplified systems all the channels lie within the EDFA gain bandwidth (~30 nm), and so fall within the Raman gain profile. Assuming N equally-spaced and equal-power channels with separation Δf, the fractional power lost by the shorter wavelength channel is:

$$D = 2.8 \times 10^{-14} N(N-1) P \Delta f L_{eff} \tag{11.19}$$

The above equation was obtained for typical system parameters of $g_r = 6 \times 10^{-14}$ m/W and fibre core cross-section area $A_{eff} = 8 \times 10^{-11}$ m^2. L_{eff} is the effective length of each fibre section in the amplified system and is defined as $L_{eff} = \{1 - \exp(-\alpha L_A)\}/\alpha$ where α is the fibre loss coefficient and L_a is the amplifier spacing. Using the theory in References 18 and 19, to ensure a SNR degradation of less than 0.5 dB in the worst channel, the maximum transmission distance in the presence of amplifier noise can be expressed as:

$$L_{max} = \left[\frac{8.7 \times 10^{15}}{2(SNR)h\nu\gamma\Delta f N(N-1)(e^{\alpha L_A}-1)L_{eff}/L_a^2} \right]^{1/2} \tag{11.20}$$

where hν, γ and Δf are the photon energy, amplifier population inversion parameter and optical bandwidth, respectively. Although this expression has been derived for the case of intensity modulated digital transmission, it can also be used to obtain an estimate of the limitation imposed on amplitude modulated analogue transmission. Figure 11.11 shows the maximum distance represented by this equation for the case where the optical filter bandwidth is 10 GHz, the bandwidth occupied by each channel is 2.5 GHz and the required optical *SNR* (before the detection process) is 30 dB - a typical *SNR* value for some AM systems (see example in Section 11.6).

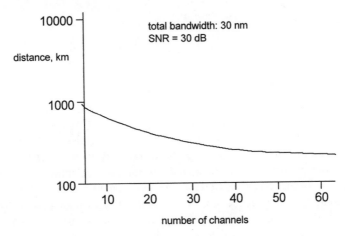

Figure 11.11 *Limitations due to stimulated Raman scattering*

For 30 channels spread over the erbium amplifier window, the maximum distance is approximately 300 km before SRS effects are noticeable. The expression shows that for a given number of channels the maximum distance is inversely proportional to the square root of the total bandwidth of the WDM signal. This means that the maximum transmission distance can be increased by using narrow channel spacing. However, narrowing channel spacing will increase the effect of four wave mixing, which is analysed in the next section.

Four wave mixing. Four wave mixing arises from the dependence of the fibre refractive index on the intensity of the optical wave propagating along the fibre [22]. In this interaction, new optical waves of frequencies f_{ijk} are generated through mixing three waves of frequencies f_i, f_j and f_k obeying the relationship f_{ijk} $= f_i + f_j - f_k$ which is required by the energy conservation condition. These newly generated waves (effectively intermodulation products) can cause crosstalk between channels in a multi-wavelength transmission system if the channels are equally spaced.

The efficiency of the FWM process is dependent on the degree of optical phase matching, which is in turn dependent on fibre dispersion and channel spacing. For this reason FWM is generally only a problem in systems operating close to the fibre dispersion zero (e.g. at 1.3 μm or at 1.5 μm in dispersion shifted fibre) with wavelength separations less than 1 nm. In the multi-wavelength transmission system of Figure 11.12, with M-1 in-line optical amplifier repeaters, the total FWM power generated at frequency f_n in the worst case (when all the generated FWM waves in each fibre segment add in phase at the end of the link) can be expressed as:

$$P_{ijk} = \kappa^2 P_s^3 e^{-\alpha L_A} \left(\frac{L_{eff}}{A_{eff}}\right)^2 \sum_{f_n = f_i + f_j - f_k} \eta_{ijk} d_{ijk}^2 \qquad (11.21)$$

where

$$\kappa = \left(\frac{32\pi^3 \chi_{1111}}{n^2 \lambda c}\right)$$

In the above equations, χ_{1111} is the third-order non-linear electric susceptibility, n is the refractive index of the fibre, λ is the wavelength, c is the velocity of light, P_s is the input power of each channel, L_A is the physical fibre length between repeaters, L_{eff} is the effective fibre length between repeaters, A_{eff} is the fibre effective area, d_{ijk} is the degeneracy factor which takes a value of 3 (for $i = j$) or 6 (for $i \neq j$) and η_{ijk} is the FWM efficiency, which can be expressed in terms of phase mismatching $\Delta\beta_{ijk}$ as:

$$\eta_{ijk} = \left(\frac{\alpha^2}{\alpha^2 + \Delta\beta_{ijk}^2}\right)\left[1 + \frac{4e^{-\alpha L_A}}{(1-e^{-\alpha L_A})^2}\sin^2(\Delta\beta_{ijk}L_A/2)\right] \qquad (11.22)$$

where α is the fibre loss coefficient. In a WDM system this FWM power will appear as noise or crosstalk in the wavelength channel located at frequency f_n. If the required signal to crosstalk noise ratio is $SNR_{FWM} = P_s\exp(-\alpha L)/P_n$ at the receiver, then the maximum allowable power per channel P_{smax} can be expressed as:

$$P_{s\max} = \min_{n=1..N}\left[\frac{A_{eff}}{ML_{eff}}\sqrt{\frac{1}{(SNR)_{FWM}\kappa^2 \displaystyle\sum_{f_n=f_i+f_j-f_k}\eta_{ijk}d_{ijk}^2}}\right] \qquad (11.23)$$

The maximum allowable power per channel is decided by the worst affected channel, and therefore the minimum value is taken above. Taking into account amplifier noise, FWM imposes a maximum distance limit given by:

$$L = \left[\frac{A_{eff}L_A^2 P_{s\max}}{2(SNR)h\upsilon\gamma\Delta f(e^{\alpha L_A}-1)L_{eff}}\right]^{1/2} \qquad (11.24)$$

Figure 11.12 shows results for the maximum distance obtained using the above expression. A dispersion coefficient of 2 ps/(nm.km) is assumed. *SNR* and *SNR_FWM* are taken as 30 dB, the optical bandwidth as 10 GHz and the fibre loss as 0.25 dB/km.

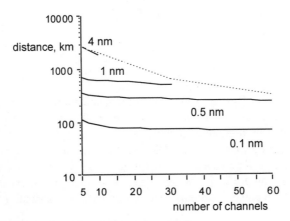

Figure 11.12 *Limitations due to four wave mixing*

The four curves show the effect for channel separations of 4nm, 1nm, 0.5 nm and 0.1 nm separation, respectively, while the dashed upper line is the boundary imposed by having a maximum total bandwidth of 30 nm. The results demonstrate that FWM becomes significant as the wavelength separation decreases. For example, with 10 channels compressed into 1 nm, i.e. approximately 0.1 nm separation, the maximum distance is less than 100 km before FWM effects become significant.

11.6 Application of amplifiers in fibre distribution systems

It is instructive to consider an example of the application of optical amplifiers to analogue TV distribution systems. Figure 11.13 shows a system using subcarrier modulation with direct fibre connection to the subscriber. At the head end the individual TV channels are assembled in a frequency division multiplex using local oscillators. The combined signal is then used to intensity modulate a laser with highly linear characteristics to avoid intermodulation distortion. In addition, as the laser power is shared by all the channels, each channel can, at maximum, have a modulation index of $1/N$, where N is the number of channels. Thus for large numbers of channels the power budget available may be very low.

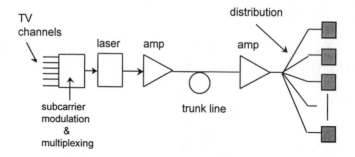

Figure 11.13 *Distribution system using optical amplifiers*

At the head-end an EDFA is used to boost the launch power and compensate for the losses associated with the trunk link, whose length may be in the order of 10-20 km. At the distribution point, at street level for example, an additional EDFA is used to enhance the splitting ratio N. The distribution line length might be in the order of 2 -5 km. The value of the splitting ratio N can then be determined as a function of the amplifier parameters [5]. It is clear from the diagram that amplifier saturation power is an important attribute of the distribution amplifier in this context.

Following the analysis of Reference 5 the carrier-to-noise ratio (*CNR*) of a single video channel at any of the optical receivers is given by :

$$CNR = \frac{m^2 I_p^2}{2\sigma^2}$$

(11.25)

where

$$I_p = \text{photodiode current} = RP_{in}$$
$$R = \text{diode responsivity}$$
$$P_{in} = \text{average received power}$$
$$\sigma^2 = \text{total mean square noise}$$

The total mean square noise is defined as:

$$\sigma^2 = \sigma_{amp}^2 + \sigma_{rec}^2 + \sigma_{rin}^2 + \sigma_{im}^2$$

where the terms on the right represent the noise components associated with the amplified signal, the thermal noise associated with the receiver, source relative intensity noise and intermodulation noise, respectively. Using the noise terms defined in Section 11.3:

$$\sigma_{amp}^2 = 2e\left[GP_{in} + P_{ase} + \gamma(G-1)P_{ase} + 2P_{in}\gamma G(G-1)\right]RB$$
$$\sigma_{rec}^2 = I_{rec}^2 B$$
$$\sigma_{rin}^2 = I_{rin}^2 B$$
$$\sigma_{im}^2 = I_{im}^2 B$$

(11.26)

where I_{rec}^2 is the receiver thermal noise spectral density, I_{rin}^2 is the current spectral density due to laser relative intensity noise and I_{im}^2 is the spectral density due to intermodulation distortion.

The average signal power per channel is:

$$P_c = m^2 G^2 P_{in}^2$$

The above equations allow the CNR to be expressed as a function of system and amplifier parameters, while the SNR required depends on the transmission signal format. Two types are generally considered for analogue TV systems, amplitude modulated vestigial-sideband (AM) and frequency modulated (FM). AM-FDM has a very poor loss budget (in the order of 8 dB for 40 channels), because it lacks

310 *Optical fibre amplifiers and WDM*

the bandwidth expansion advantage of FM-FDM, however it is attractive from a customer point of view as it is compatible with existing receivers. Typically for 40 channels a CNR of 55 dB is required for short-span video transmission. This requirement leads to poor effective receiver sensitivities (in the order of - 6 dBm). Owing to the low power budget associated with AM-FDM, its use in distribution systems is limited, so it is of interest to see how the performance can be enhanced through the use of optical amplifiers. In contrast, FM-FDM has a much improved power budget, since the use of bandwidth expansion reduces the required CNR to around 15 dB, giving a loss budget of approximately 27 dB for 40 channels.

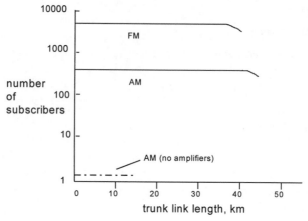

Figure 11.14 *Performance of amplified distribution system*

Figure 11.14 illustrates the improvements in distribution numbers obtained through the use of amplifiers for both AM and FM. In the calculations it is assumed that the amplifier saturation power is + 20 dBm and that fibre losses are 0.5 dB/km. The results show that significant improvements are obtained through the use of optical amplifiers. For example in the AM case the distribution number increases from single figures to approximately 300 with trunk lengths of up to 40 km possible, while FM can support well over several thousand users over similar distances.

11.7 References

1   Ainslie, B.J., Craig-Ryan, S.P., Davey, S.T., Armitage, J.R., Atkins, C.G., Massicott, J.F., and Wyatt. R.: 'Erbium doped fibres for efficient optical amplifiers', *IEE Proc.- J*, 1990, **137**, pp. 205-208
2   O' Mahony, M.J.: 'Semiconductor laser optical amplifiers for use in future fiber systems', *J. Lightwave Technol.*, 1988, **6**, pp. 531-544

3   Olsson, N.A.: 'Lightwave systems with optical amplifiers', *J. Lightwave Technol.*, 1989, **7**, pp. 1071-1082

4   Gillner, L.: 'Properties of optical switching networks with passive or active space switches', *IEE Proc.-J*, 1993, **140**, pp. 309-315

5   Nakagawa, K., Nishi, S., Aida, K., and Yoneda, E.: 'Trunk and distribution network application of erbium doped fibre amplifier', *J. Lightwave Technol.*, 1991, **9**, pp. 198-208

6   Whitley, T.J., Wyatt, R., Szebesta, D., and Davey, S.T.: 'Towards a practical 1.3 μm optical fibre amplifier', *BT Technol. J.*, 1993, **11**, pp. 115-127

7   Lipka, D.: 'Large signal models for erbium doped fibre amplifiers', *Philips Telecomm. Review*, 1993, **51**, pp. 31-40

8   Giles, C.R., and Desurvire, E.: 'Propagation of signal and noise in concatenated erbium doped fiber optical amplifiers', *J.Lightwave Technol.*, 1991, **9**, pp. 147-155

9   Agrawal, G.P.: 'Fiber-optic communication systems' (John Wiley & Sons, New York, 1992)

10   Atkins, C.G., Armitage, J.R., Wyatt, R., Ainslie, B.J., and Craig-Ryan, S.P.: 'High gain , broad spectral bandwidth erbium doped fibre amplifier pumped near 1.5 μm', *Electron. Lett.*, 1989, **25**, pp. 910-912

11   Yamamoto, Y.: 'Noise and error rate performance of semiconductor laser amplifiers in PCM-IM optical transmission systems', *IEEE J. Quantum Electron.*, **QE-16**, 1980, pp. 1073-1081

12   Senior, J.M.: 'Optical fibre communications' (Prentice Hall, London, 1992)

13   Brackett, C.A.: 'Dense wavelength division multiplexing networks: principles and applications', *IEEE J. Select. Area Comm.*, 1990, **8**, pp. 948-964

14   de Albuquerque, A.A., Houghton, A.J.N., and Malmros, S.: 'Field trials for fiber access in the EC', *IEEE Communications*, 1994, **32**, pp.40-43

15   Green, P.E.: 'Fiber optic networks' (Prentice Hall, New York, 1993)

16   Hill, G.R., Chidgey, P.J., Kaufhold, F., Lynch, T., Sahlen, O., Gustavsson, M., Janson, M., Lagerstrom, B., Grasso, G., Meli, F., Johannson, S., Ingers, J., Fernandez, L., Rotolo, S., Antonielli, A., Tebaldini, S., Vezzoni, E., Caddedu, R., Caponio, N., Testa, F., Scavennec, A., O'Mahony, M.J., Zhou, J., Yu, A., Sohler, W., Rust U., and Herrmann, H.: 'A transport Network layer based on optical network elements', *IEEE J. Lightwave Technol.*, 1993, **11**, pp. 667-680

17   Stolen, R.H.: 'Nonlinearity in fiber transmission', *Proc. IEEE*, 1980, **68**, pp 1232-1236

18   Chraplyvy, A.R.: 'Limitations on lightwave communications imposed by optical fibre nonlinearities', *IEEE J. Lightwave Technol.*, 1990, **8**, pp. 1548-1557

19   Chraplyvy, A.R.: 'What is the actual capacity of single mode fibre in amplified light systems', *IEEE Photonics Technol. Lett.*, 1993, **5**, pp. 665-668

20   Miller, C.A.: 'Characteristics and applications of high performance tunable fibre Fabry-Perot filters', ECTC Electronics Components & Technology Conference, Atlanta, May 1991

21   Willner, A.E.: 'SNR analysis of crosstalk and filtering effects in an amplified multichannel direct-detection dense-WDM system', *IEEE J. Photon. Technol. Lett.*, 1992, **4**, pp. 186-189

22   Ellis, A.D, and Stallard, W.A.: 'Four wave mixing in ultra long transmission systems incorporating linear amplifiers', IEE Colloquium Digest No. 1990/159, Savoy Place, London, November 1990

Index

Opto-electronic integrated circuit
    (OEIC) 48, 232
Optoelectronic mixers 207
Optoelectronic mixing 223
Optimisation 274

P

Passive optical network (PON) 222
Perturbation technique 184, 188
Phase modulation 46, 57, 59
Phase noise 239
Phase resolution 60
Photoconductor 52
Photodiode:
    - edge-entry 218-220
    - pin 52, 57, 218, 239, 258
    - zero bias 220, 258
Photo HBT 222, 225
Picocell 204
Polarisation:
    - control 36
    - diversity 40
Power series 181
Power spectral density 159
Praseodymium amplifier 294
Propagation equations 291
Probing method 185
Pulse analogue modulation 123, 125
Pulse analogue techniques 52, 62
Pulse code modulation 116, 140
Pulse equaliser 66
Pulse frequency modulation 52, 63,
    71-75, 92, 106-107
Pulse interval and width modulation
    63, 92, 103-105
Pulse interval modulation 63, 92,
    101-103
Pulse phase modulation 125
Pulse position modulation: 62, 70,
    91, 99-101,123, 127, 139
    - digital 91
    - direct digital 134
    - indirect digital 134

    - large deviation 130
    - linear 127
    - modulation depth 70
    - multiplexed 134
    - small deviation 130
    - uniformly sampled 133
Pulse time modulation:
    - baseband component 93
    - carrier frequency 93
    - classification 94
    - demodulation 93
    - family 90, 139
    - hybrid technique 110-111
    - improvement factor 115-116
    - noise threshold 94
    - nonlinear distortion 111-115
    - performance potential 111
    - performance trade-off 94, 95
    - pulse jitter 94,133, 111
    - sidetones 93
    - signal-to-noise ratio 90, 111-
      116
    - spectral overlap 111
    - spectrum 93
    - transmission channel
      bandwidth 115-116
Pulse to noise ratio 68, 70
Pulse width modulation 62, 63, 67-
    69, 91, 95-99

Q

Quantum confined Stark effect
    43, 212

R

Radio LANs 205
Rare-earth doped-fibre amplifier 298
Rate equations 289-290
    - normalised 184
Research for advanced
    communications in Europe
    (RACE) 205, 250